I0390608

North Africa: The Impact of Climate Change to 2030 (Selected Countries)

A Commissioned Research Report

The National Intelligence Council sponsors workshops and research with nongovernmental experts to gain knowledge and insight and to sharpen debate on critical issues. The views expressed in this report do not reflect official US Government positions.

NIC 2009-007D
August 2009

This page is intentionally kept blank.

Scope Note

Following the publication in 2008 of the National Intelligence Assessment on the National Security Implications of Global Climate Change to 2030, the National Intelligence Council (NIC) embarked on a research effort to explore in greater detail the national security implications of climate change in six countries/regions of the world: India, China, Russia, North Africa, Mexico and the Caribbean, and Southeast Asia and the Pacific Island states. For each country/region, we are adopting a three-phase approach.

- In the first phase, contracted research—such as this publication—explores the latest scientific findings on the impact of climate change in the specific region/country.

- In the second phase, a workshop or conference composed of experts from outside the Intelligence Community (IC) will determine whether anticipated changes from the effects of climate change will force inter- and intra-state migrations, cause economic hardship, or result in increased social tensions or state instability within the country/region.

- In the final phase, the NIC's Long-Range Analysis Unit (LRAU) will lead an IC effort to identify and summarize for the policy community the anticipated impact on US national security.

To support research by the National Intelligence Council on the national security impacts of global climate change, this assessment of the impact of climate change on North Africa through 2030 is being delivered under the Global Climate Change Research Program contract with the Central Intelligence Agency's Office of the Chief Scientist.

This assessment identifies and summarizes the latest peer-reviewed research related to the effects of climate change on selected countries in North Africa, drawing on the literature summarized in the latest Intergovernmental Panel on Climate Change (IPCC) assessment reports, National Communications to the United Nations Framework (UNFCCC) on Climate Change, and on other peer-reviewed research literature and relevant reporting. It includes such impacts as sea-level rise, water availability, agricultural shifts, ecological disruptions and species extinctions, infrastructure at risk from extreme weather events (severity and frequency), and disease patterns. This paper addresses the extent to which the countries in the region are vulnerable to climate change impact. The targeted time frame is to 2030, although various studies referenced in this report have diverse time frames.

This assessment also identifies (Annex B) deficiencies in climate change data that would enhance the IC's understanding of potential impacts on North Africa and other countries/regions.

This page is intentionally kept blank.

Executive Summary

Model projections available for North Africa indicate a clear increase in temperature over the next 20 years that is expected to continue throughout the 21st century, probably at a rate higher than the estimated global average. Model simulations also suggest a drying trend in the region, particularly along the Mediterranean coast, driven by large decreases expected in summertime precipitation. Because coastal areas historically receive by far the largest amount of rainfall in North Africa, future decreases will likely have a significant and noticeable impact. Precipitation trends in the interior semiarid and arid regions of North Africa are more difficult to predict due to the very small amount of natural precipitation that characterizes these areas. Climate change will induce some variations in precipitation patterns, but the trend is not clear, as some models predict slight increases and others predict slight decreases in annual precipitation amounts.

The Regional Climate Change Index (RCCI)[1] identifies the Mediterranean as a very responsive region to climate change ("Hot-Spot"). Given the ecological and socioeconomic characteristics of the southern Mediterranean countries, the impact of climate change may be more marked than in other regions of the world. Still, most of the predicted impacts in the region are already occurring regardless of climate change (e.g., water stress and desertification). Climate change is expected to exacerbate these trends.

Based on global climate projections and given inherent uncertainties, the most significant impacts of climate change in North Africa (Morocco, Algeria, Tunisia, Libya, and Egypt) will likely include the following:

- *Water Resources Stress.[2]* All countries of North Africa are presently experiencing water stress. Model simulations show a general decrease in rainfall across North Africa, with median decreases in average annual precipitation of 12 percent and 6 percent projected for the Mediterranean and Saharan regions, respectively. This general drying trend for North Africa is punctuated by seasonal variations in projected precipitation that differ by region. Predicted decreases in average annual rainfall, accompanied by projected increases in the population of the region, may impede access to water for millions of inhabitants. In addition, with decreasing water levels, other ecological effects such as salinity in coastal areas and deterioration of water quality may increase.

- *Agriculture.* Model results are inconsistent regarding future changes in crop yields and agricultural growing seasons in North Africa, and we do not know whether variations in temperature, precipitation, or atmospheric CO_2 will be the dominant factor. One modeling study suggests that future increases in atmospheric CO_2 concentrations will increase maize yields in Morocco, while other modeling studies suggest that future increases in air temperature will have a negative effect on growing seasons and crop yields in Egypt. Relatively heat-tolerant species, such as maize, are expected to suffer the smallest losses in yield and growing area, while heat-intolerant crops, such as wheat, are expected to suffer the largest losses. In addition, intensive irrigation practices in the region may result in further

---

[1] The RCCI is calculated for 26 land regions from projections of 20 global climate models using the Intergovernmental Panel on Climate Change (IPCC) emission scenarios.
[2] Water Stress, as used by the IPCC, refers to a per capita water availability of below 1,000 cubic meters per person per year; sometimes IPCC referenced sources also use a ratio of withdrawals to long-term average runoff of 0.4. The IPCC formally defines a country as water stressed when withdrawals exceed 20% of renewable water supply.

salinity, which may lead to desertification. Adaptation strategies, including modifications in sowing dates to match climate changes and development of heat-tolerant crop varieties, will likely mitigate some of the expected negative effects on North African agriculture. Development of regional and local climate models in the coming years that include projections of Mediterranean Sea level rise and decreases in the Nile River flow are expected to provide more accurate estimates of future changes in North African agricultural regions.

- *Migration.* In recent years, North Africa has experienced vast migration pressures from both migrants that settle in the region from the south or that use North African countries as a transit area to reach Europe. Thus far, experts have not cited climate change as a driving force for migration in the region; nevertheless, a warmer climate and changing precipitation patterns, which will likely reduce viable cropland and reduce access to water, will increase urbanization and make accommodating the needs of a growing population more difficult. Besides food and water necessities, climate change-related migration may also imply greater demands on infrastructure along the coasts as well as ethnic, racial, or religious clashes.

- *Natural Disasters.* Because of the lack of historical data from tide gauges in the region, the wide range of future estimates in sea level, and the paucity of regional climate model projections for the Mediterranean Sea, a definitive estimate of sea level rise along the coastline of North Africa in the next 20 years is not possible. However, the intensity and frequency of floods along the Mediterranean coast are expected to increase by the middle of the 21st century. Compared to other regions, the impacts of sea level rise in North Africa are expected to be stronger in terms of social, economic, and ecological factors. Highly populated and agriculturally important coastal cities are the most vulnerable.

In addition, two more potentially serious impacts are the following:

- *Tourism.* Tourism is an important source of income for most countries of North Africa. Of concern, however, are the large quantities of water this sector demands and the little attention that governments of this region have given to water provision in the past. Thus, increased water scarcity, sea level rise, and increasing temperatures will likely have a negative impact on this sector and consequently the economy of most North African countries.

- *Energy.* The economies of Algeria, Libya and (to a lesser extent) Egypt are dependent on the hydrocarbon industry. Because of the revenues they receive from exporting fossil fuels—mostly to Europe—they are to some degree more resilient to the deleterious impacts of climate change. Any shift in the interest of other regions in importing natural gas and oil from North Africa, conversely, may make these North African countries considerably more vulnerable. However, there is no indication now that Europe and other importing regions will stop importing from North Africa in the next few decades.

Based on a comprehensive global comparative study of resilience to climate change (including adaptive capacity) using the Vulnerability-Resilience Indicators Model, a wide range of adaptive capacity is represented in this group of countries from Libya (ranking 34th in a 160-country study) to Morocco (ranking 136th in the same study). Under a high-growth scenario of the future, all countries gain adaptive capacity, especially Libya. However, under a delayed-growth scenario, all of these North African countries lose adaptive capacity.

Contents

This page is intentionally kept blank.

Introduction and Background

North Africa is vulnerable to climate change impacts, both direct and indirectly, e.g., as a result of any actions to reduce greenhouse gas emissions and thus consumption of fossil fuels. This report summarizes peer-reviewed and other relevant research about this region, including projected climate changes, impacts on human and natural systems, and the adaptive capacities of countries in the region. Literature sources include the 2007 Intergovernmental Panel on Climate Change (IPCC) Assessment Report, peer-reviewed journal articles, and reports generated by governments and scientific organizations.

The report focuses on the five northernmost countries in Africa: Morocco, Algeria, Tunisia, Libya, and Egypt. Collectively, the report refers to these countries as "North Africa," although we recognize that other groups of countries are given this same label.[3] All five countries are coastal, with northern borders on the Mediterranean Sea. A brief description of each country is given below.

Morocco
The Kingdom of Morocco borders the Mediterranean Sea to the north, the North Atlantic Ocean to the west (the two bodies of water separated by the Straits of Gibraltar), Western Sahara to the south, and Algeria to the south and east. Morocco's area totals 446,550 sq km, including 250 sq km of water. Its northern coast and interior are mountainous, but it also has plateaus, valleys, and rich coastal plains. The northern mountains are subject to earthquakes, and the country experiences periodic droughts. Current environmental issues include soil erosion from degradation-causing land management, water polluted by raw sewage, siltation of reservoirs, and oil pollution of coastal waters. Its population is about 35 million (2009 estimate). Life expectancy at birth is 72 years. Its people are 99 percent Muslim, with the rest Christian or Jewish. Gross Domestic Product (GDP) per capita in 2008 was estimated by the CIA at $4,000 United States dollars (USD) equivalent, with an unemployment rate of 10 percent (2008 estimate).

Algeria
The People's Democratic Republic of Algeria is bordered on the west by Morocco, Western Sahara, and Mauritania; on the southwest by Mali; on the southeast by Niger; and on the east by Libya and Tunisia. Its area totals 2,381,740 sq km, with no areas of water. Algeria's terrain is mostly high plateau and desert, with some mountains and a narrow, discontinuous coastal plain. Its mountainous areas experience severe earthquakes, and the country is subject to mudslides and floods in rainy seasons. Algerian population totals about 34 million (2009 CIA estimate). Life expectancy at birth is 74 years. The population is 99 percent Sunni Muslim (the state religion), with Christians and Jews composing the remainder. Oil and gas account for approximately 60 percent of budget revenues, 30 percent of GDP, and more than 95 percent of export earnings. GDP per capita was estimated for 2008 at $7,000 USD equivalent, with an unemployment rate of 13 percent (2008 estimate).

---

[3] For a map of North Africa, demarcated by the red line, see: Welt-atlas.de: Atlas of the World, s.v. "Map of Africa, North," http://www.welt-atlas.de/datenbank/karten/karte-0-9008.gif (accessed May 11, 2009).

Tunisia

The Tunisian Republic is located between Algeria and Libya. Its area totals 163,610 sq km, including 8,250 sq km of water. Tunisia has mountains in the north; a hot, dry central plain; and semiarid regions in the south (merging into the Sahara Desert). Current environmental issues include ineffective toxic and hazardous waste disposal, water polluted by raw sewage, limited natural fresh water resources, deforestation, overgrazing, and soil erosion. Tunisia's population totals about 10.5 million (2009 estimate). Life expectancy at birth is 76 years. Tunisians are 98 percent Muslim, 1 percent Christian, and 1 percent Jewish and other religions. GDP per capita in 2008 was estimated by the CIA at $7,900 USD equivalent, with a 14 percent unemployment rate (2008 estimate).

Libya

The Great Socialist People's Libyan Arab Jamahiriya lies between Egypt and Tunisia, also bordering Sudan to the southeast, Niger and Chad to the south, and Algeria to the west. Libya's area totals 1,759,540 sq km, including no areas of water. Its terrain is flat to undulating plains, plateaus, and depressions; its landscape is mostly barren. The country is subject to dust and sand storms; the ghibli, a southern wind that lasts one to four days, occurs in spring and fall. Current environmental issues include desertification and extremely limited fresh natural water. Total population is approximately 6.3 million (2009 CIA estimate). Life expectancy at birth is 77 years. Citizens are 97 percent Sunni Muslim. Revenues from oil contribute about 95 percent of export earnings, about one-quarter of GDP, and 60 percent of public sector wages; Libya imports about three-quarters of its food. GDP per capita in 2008 was estimated by the CIA at $14,400 USD equivalent; the unemployment rate for 2004 was estimated at 30 percent.[i]

Egypt

The Arab Republic of Egypt is located between Libya to the west and the Gaza Strip and the Red Sea to the east; Egypt includes the Asian Sinai Peninsula and the Suez Canal. Egypt's area totals 1,001,450 sq km, including 6,000 sq km of water. Except for the Nile River valley and delta, Egypt is a desert plateau. It is subject to periodic droughts, earthquakes, flash floods, landslides, dust and sand storms, and a driving windstorm called khamsin in the spring. Current environmental issues encompass loss of agricultural land to urbanization, soil salination below the Aswan High Dam; oil and water pollution, and limited natural fresh water that is increasingly stressed by population growth. Current population is about 83 million (2009 estimate). Life expectancy at birth is 72 years. Citizens are 90 percent Muslim (mostly Sunni), 9 percent Coptic, and 1 percent Christian and other religions. GDP per capita in 2008 was estimated by the CIA to be $5,400 USD equivalent, with an unemployment rate of 8.7 percent.

Projected Regional Climate Change

Current Climatology of North Africa

The climate of North Africa varies substantially between coastal and inland areas of the region. Along the coast, North Africa has a Mediterranean climate, which is characterized by mild, wet winters and warm, dry summers, with ample rainfall of approximately 400 to 600 mm per year. Inland, the countries of North Africa have semiarid and arid desert climates, which are marked by extremes in daily high and low temperatures, with hot summers and cold winters, and little rainfall—approximately 200 to 400 mm per year for semiarid regions and less than 100 mm per year for desert regions.[ii] Figure 1, a plot of average daily rainfall in North Africa for the period

1983-2005, illustrates the demarcation between the coastal and inland climate zones. The Mediterranean climate zone, indicated by the blue-green and turquoise-colored shading, runs along a relatively thin strip of land bordering the coasts of Morocco, Algeria, Tunisia, and parts of Libya and Egypt. The semiarid climate zone, indicated by the blue shading, is a transition zone between the Mediterranean zone and the arid desert climate zone, which is indicated by the dark blue shading. As Figure 1 shows, the semiarid and desert climate zones dominate, and most of North Africa is very dry. Along the coast, the rainy season typically runs from October to March or April. Torrential downpours during the rainy season can cause devastating flooding, and droughts occur frequently in the dry inland regions, sometimes lasting for years at a time.[iii] Figures 2 through 4 show the monthly average daily minimum temperatures, daily maximum temperatures, and total rainfall amounts for the capital cities in North Africa, all of which are situated on the Mediterranean coast and thus have ample rainfall, except for Cairo, Egypt.

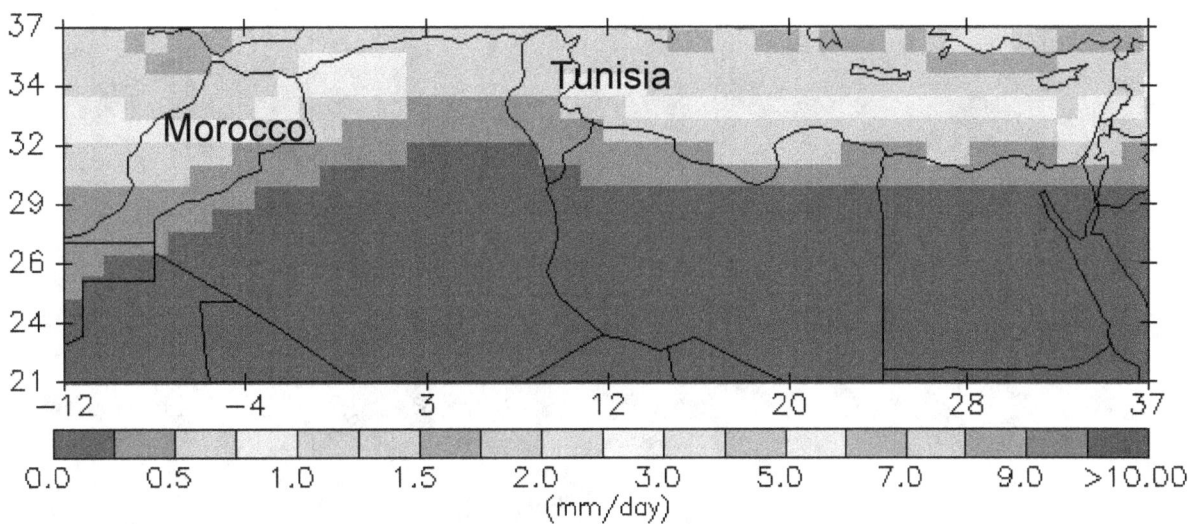

Figure 1. Daily average precipitation rates for North Africa during the period 1983-2005. Latitude (°N) is listed on the y-axis and longitude is listed on the x-axis (negative values are °W and positive values are °E). Precipitation values were measured by satellite; daily values were derived by dividing the total monthly averaged amount of precipitation for a given month by the number of days in the month for the 22-year climatology period. Source: NASA Atmospheric Science Data Center, *NASA Surface Meteorology and Solar Energy: Global/Regional Data,* http://eosweb.larc nasa.gov/cgi-bin/sse/sse.cgi?na+s01#s01 (accessed May 18, 2009).

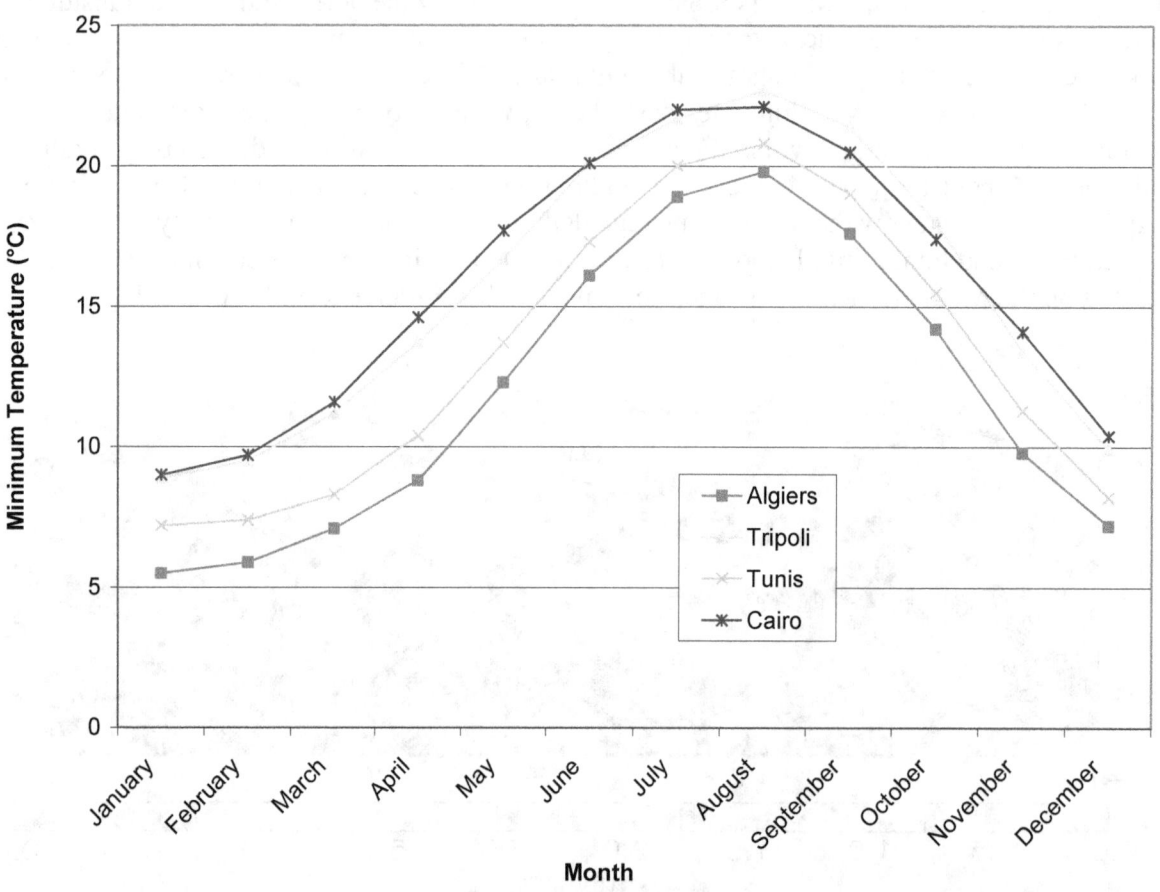

Figure 2. Monthly averaged daily minimum temperatures for capital cities in North Africa. Climatology values for Algiers, Algeria are averaged over the period 1976-2005; values for Tunis, Tunisia are averaged over the period 1961-1990; values for Tripoli, Libya are averaged over the period 1961-1990; and values for Cairo, Egypt are averaged over the period 1971-2000; values for Rabat, Morocco are not available. *Source*: World Meteorological Organization, *World Weather Information Service*, http://www.worldweather.org/ (accessed May 15, 2009).

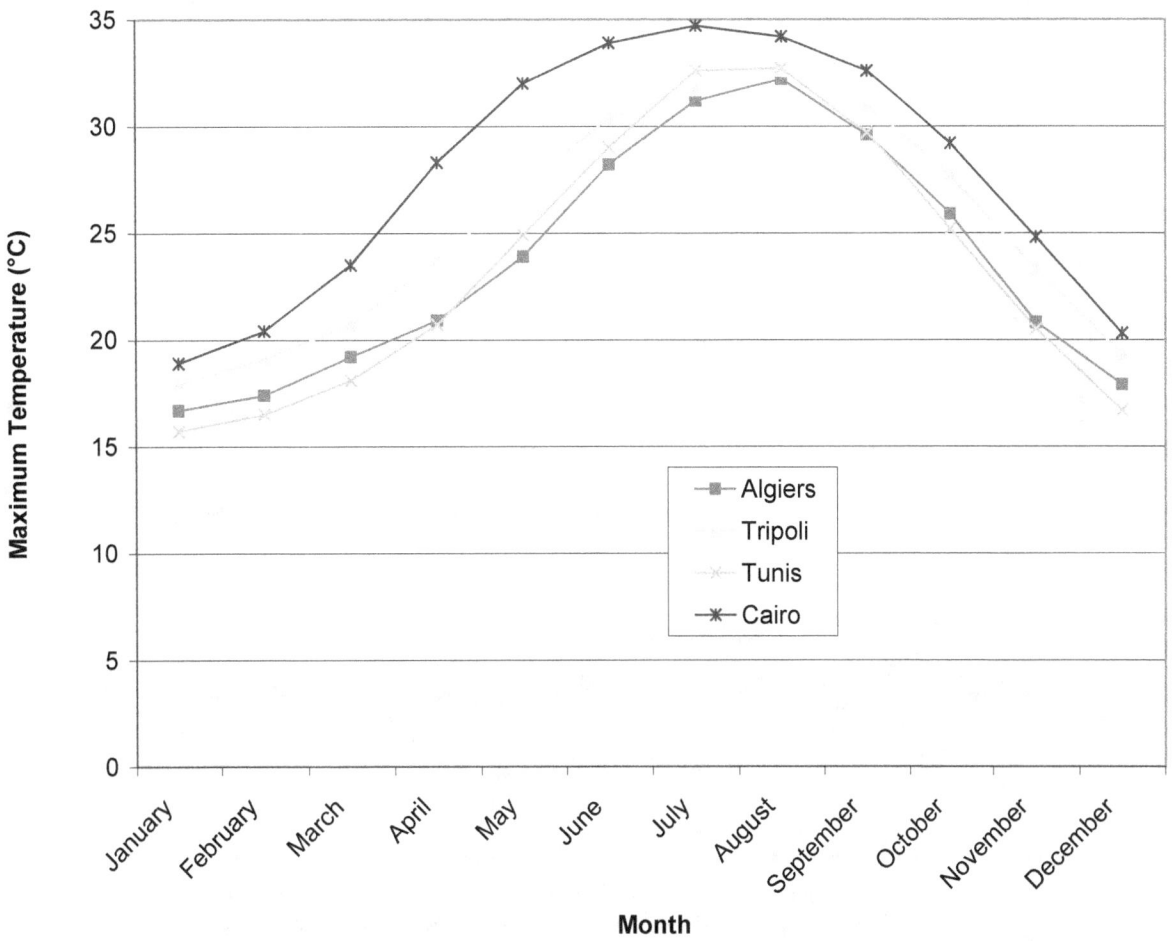

Figure 3. Monthly averaged daily maximum temperatures for capital cities in North Africa. Climatology values for Algiers, Algeria are averaged over the period 1976-2005; values for Tunis, Tunisia are averaged over the period 1961-1990; values for Tripoli, Libya are averaged over the period 1961-1990; and values for Cairo, Egypt are averaged over the period 1971-2000; values for Rabat, Morocco are not available. Source: World Meteorological Organization, *World Weather Information Service*, http://www.worldweather.org/ (accessed May 15, 2009).

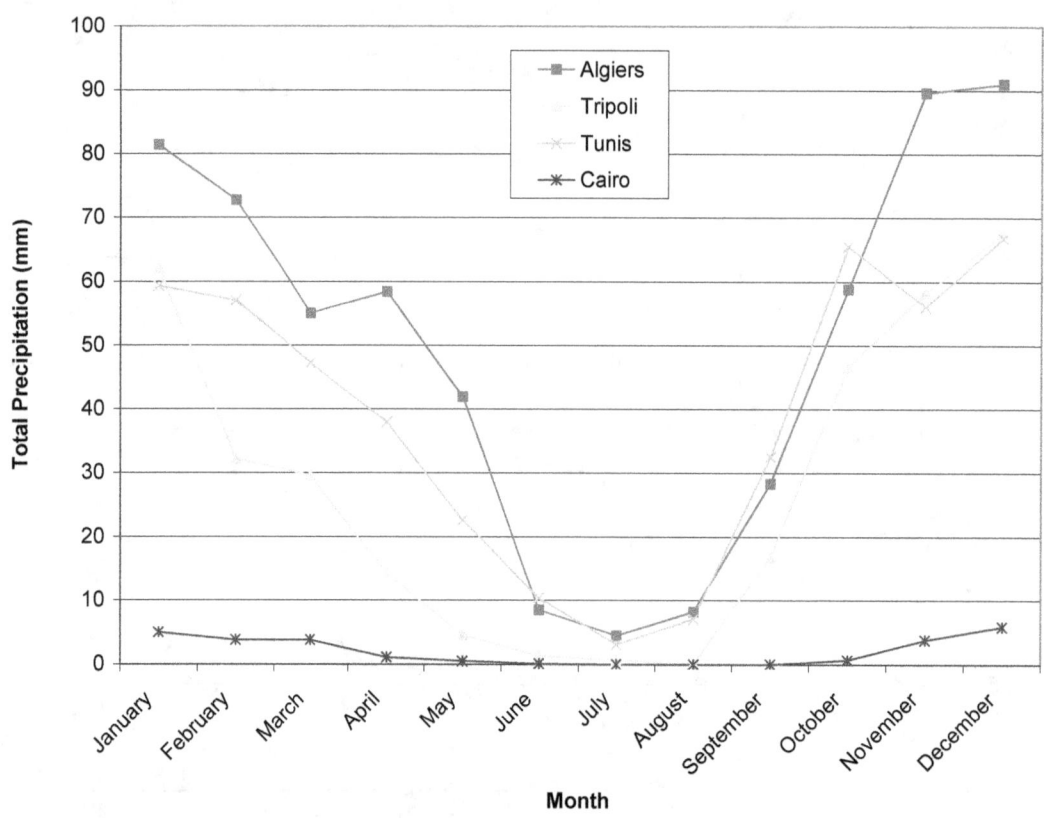

Figure 4. Monthly averaged total precipitation values for capital cities in North Africa. Climatology values for Algiers, Algeria are averaged over the period 1976-2005; values for Tunis, Tunisia are averaged over the period 1961-1990; values for Tripoli, Libya are averaged over the period 1961-1990; and values for Cairo, Egypt are averaged over the period 1971-2000; values for Rabat, Morocco are not available. Source: World Meteorological Organization, *World Weather Information Service*, http://www.worldweather.org/ (accessed May 15, 2009).

An important distinctive climatic feature of North Africa is the *sirocco*, a hot, dry, southerly wind that occurs year round. Siroccos are typically strongest in the spring, when they can reach gale force strength (defined as sustained wind speeds of 34 to 47 knots). Wind events generally last 10 to 12 hours, but events as long as 36 hours have been observed. The winds originate over the Sahara Desert and blow north across North Africa, over the Mediterranean Sea, and into southern Europe. Because siroccos flow from the desert, they typically contain large amounts of sand and dust that limit visibility and can damage machinery. Siroccos are caused by the west-to-east progression of extratropical cyclones (low pressure systems) across the Mediterranean.[4]

---

[4] Sirocco winds are caused by the eastward progression of extratropical cyclones across the Mediterranean Sea where a low pressure center entrains southerly winds in a warm sector and thereby generates the sirocco winds. See: Weather Online, *Wind of the World: Sorocco*, http://www.weatheronline.co.uk/reports/wind/The-Sirocco htm (accessed May 15, 2009).

The sirocco is called by different names across the North African region, including *chili* in Tunisia, *ghibli* in Libya, and *khamsin* in Egypt.[iv]

Climate Predictions (Modeling)

General circulation models (GCMs) are the main tool used by scientists to project future climate change. These models simulate atmospheric and oceanic circulations, as well as processes that occur on land. As a result, GCMs are very complex models, and they tend to have rather low spatial resolutions, on the order of 400 to 125 km. To obtain model information on the local and regional scales, such as for North Africa, at higher resolutions than native GCM grid sizes, "downscaling" is used. There are two main downscaling methods, dynamical and statistical. Dynamical downscaling involves the use of high-resolution climate models with observed or simulated data as boundary conditions. This approach has high credibility, but it is computationally expensive. In contrast, statistical downscaling, which involves application of established relationships between observed data and modeled data, is computationally inexpensive, and it can replicate finer scales than dynamical downscaling. Statistical downscaling methods do not accurately simulate regional feedback effects, however.[v]

In general, GCM predictions of temperature changes for a given region are consistent, but predictions of precipitation changes can vary widely due to the difficulty in simulating the myriad factors that influence precipitation frequency, duration, and intensity. An additional complication for North Africa is the fact that most of the region has an arid or semiarid climate and thus receives little rainfall annually, as shown in Figure 1. Given the very low level of annual rainfall that occurs in the region, it is inherently difficult to predict changes in precipitation associated with future climate change. A more detailed discussion of the ability of GCMs to project regional climate changes is given in Annex A.

GCMs simulate changes in climate under scenarios of future greenhouse gas and aerosol emissions. The 2000 IPCC *Special Report on Emission Scenarios* (SRES)[vi] laid out the four basic scenario families used by IPCC scientists to predict future climate change; they are summarized in Table 1. This set of scenarios is designed to represent the range of possible future global conditions that will influence greenhouse gas emissions. The scenarios are based on consistent and reproducible assumptions about global forces that impact greenhouse gas emissions, including economic development, population, and technological change.

Climate researchers frequently use GCMs from the UK Met Office Hadley Centre for Climate Prediction and Research to investigate future changes in temperature and precipitation. These models are representative of many GCMs used to simulate the effects of climate change. The HadCM2 model has four different integrations that represent the climate effects of greenhouse gases and sulfate aerosols. Greenhouse gases, such as carbon dioxide, water vapor, ozone, methane, and nitrous oxide, absorb infrared radiation emitted from the Earth and subsequently emit it back into the atmosphere, which results in a net warming of the Earth's surface. HadCM2 includes the combined forcing of all greenhouse gases as an equivalent CO_2 concentration of 0.5 percent or 1 percent, depending on the integration. HadCM2 can also incorporate the negative direct forcing of sulfate aerosols by means of an increase in clear-sky albedo; sulfate forcing is 0.5 percent or 1 percent, depending on the model integration. The influence of sulfate aerosols is important because they reflect incoming solar radiation; thus less energy reaches the surface of the Earth, which results in a net cooling of the Earth's surface. Each integration of HadCM2 has four ensembles, from which an ensemble mean can be calculated.[vii] Ensembles are used to represent the range in uncertainty of model predictions. In this case, the same model, HadCM2, is run four times using different initial conditions. The average of a series of ensembles is always more accurate than any single model run. HadCM2 has a spatial resolution of 2.5° latitude by 3.75° longitude. In general, this resolution is sufficient to resolve climate changes on a country-level scale in North Africa, without the need for downscaling or temporal smoothing.[viii] To simulate local climate changes, however, downscaling is required.

Emission Scenario	Economic Development	Global Population	Technology Changes	Theme
A1	Very rapid	Peaks around mid-21st century and declines thereafter	Rapid introduction of new and more efficient technologies	Convergence among regions; increased cultural and social interactions
A2	Regionally-oriented	Continuously increasing	Slower and more fragmented than A1, B1, and B2	Self-reliance and preservation of local identities
B1	Rapid change toward service and information economy	Same as A1	Introduction of clean and resource-efficient technologies	Global solutions to economic, social, and environmental sustainability
B2	Intermediate levels of economic development	Continuously increasing, but not as fast as A2	Less rapid and more diverse changes than A1 and B1	Local solutions to economic, social, and environmental sustainability

Table 1. Summary of IPCC emissions scenarios. Source: Intergovernmental Panel on Climate Change (IPCC), *Special Report on Emissions Scenarios (SRES)*, eds. Nebojsa Nakicenovic and Rob Swart (Cambridge: Cambridge University Press, 2000), http://www.ipcc.ch/ipccreports/sres/emission/index htm.

In contrast to the most recent GCMs, which are run under conditions matching the various IPCC emissions scenarios, many GCMs prior to approximately 2000 were run under more simplistic conditions. The most common method of simulating climate change in the older models was with an equivalent doubling of atmospheric CO_2 concentrations ($2\times CO_2$), which represented the net radiative effect of increases in CO_2 and other greenhouse gases since pre-industrial times (typically equivalent to 560 ppm of CO_2). Models established a baseline using "current" CO_2 concentrations ($1\times CO_2$), and the change between $1\times CO_2$ and $2\times CO_2$ in model output was considered representative of future climate change. Under this type of scenario, researchers often neglected to frame the model results in terms of specific decadal changes, so the exact timeframe for projected climate changes was not specified.

Additional information on the GCMs mentioned in this report is available from the IPCC Data Distribution Centre (http://www.ipcc-data.org/).

Projections of Future Changes in Temperature and Precipitation
According to M Humle et al. average annual surface temperatures and, to a lesser degree, total precipitation amounts have changed in North Africa during the 20th century.[ix] Surface temperatures have risen across the region, especially in Tunisia and northern Algeria, which experienced an approximately 2 to 3°C increase in temperature. The trend in precipitation is less certain, likely due to the relatively small amounts of rainfall historically observed across much of

North Africa. There was a slight increase of 0 to 10 percent in total rainfall observed in the semiarid and arid regions and a slight decrease of 0 to 10 percent observed along the coastal regions during the 20th century.

We do not know the degree to which these observed trends in temperature and precipitation are due to the influence of climate change versus other anthropogenic effects, particularly changes in land cover and land use. Several researchers[x] have suggested that soil degradation, vegetation loss, and deforestation in Africa associated with changes in agricultural and grazing practices, urbanization, and construction of transportation infrastructure over the past 50 years have been major drivers for regional temperature and precipitation variability, and that increases in greenhouse gas and aerosol concentrations have played a lesser role.

Rising concentrations of greenhouse gases in the global atmosphere during the 21st century are expected to cause a net warming of the Earth's surface, and the relative influence of changes in land use and cover in Africa may contribute to this trend as well. A recent modeling study[xi] that simulated future changes in temperature and precipitation in Africa due to greenhouse forcing and land use changes found that the effect of land cover on total climate change appeared to be limited to tropical Africa and did not influence adjacent regions, such as the Sahara Desert or Mediterranean Basin. As a result, changes in land use and cover are likely to have a limited effect on future changes in temperature and precipitation in North Africa; the largest influence probably will be due to forcing from greenhouse gases and aerosols.

To quantify regional future changes in temperature and precipitation associated with climate change, the IPCC uses a coordinated set of climate model simulations archived at the Program for Climate Model Diagnosis and Intercomparison (called the multi-model dataset, or MMD). Although the MMD models have significant systematic errors in prediction of some major observed climatic events in Africa, such as rainfall in southern Africa, placement of the Atlantic Inter-Tropical Convergence Zone (ITCZ), and ocean upwelling off the West Africa coast, they have predicted robust regional trends in temperature and precipitation for North Africa.[xii] Figure 5 shows the projected increases in temperature for the Southern European and Mediterranean (SEM) and Saharan Africa (SAH) regions for the 21st century. As defined by the IPCC and shown by the green highlighted areas in Figure 5, the Southern European and Mediterranean region encompasses 30°N to 48°N latitude and 10°W to 40°E longitude, and the Saharan Africa region encompasses 18°N to 30°N latitude and 20°E to 65°E longitude. Taken together, the southernmost portion of the SEM region combined with the western two-thirds of the SAH region comprise the North African area that is the focus of this report. The SEM region corresponds to the coastal areas of North Africa that have a Mediterranean climate, while the SAH region corresponds to the inland areas of North Africa that have semiarid and arid desert climates.

To obtain the temperature information shown in Figure 5, a subset of 58 simulations from 14 models of the MMD was used for the observed period and 47 simulations from 18 models for the future projections; the future projections were calculated for the A1B emissions scenario. The width of the shading and the bars in Figure 5 represent the 5 to 95 percent range of the model output. Model simulations are presented in the context of observed warming during the 20th century, which is important because if GCMs cannot accurately reproduce observed climatic data, they cannot be relied upon to simulate future climate changes. Numerical results from simulations of 21 MMD models are summarized in Table 2 and show that by the end of the 21st century, annual mean temperatures in the Mediterranean and Saharan regions of North Africa are expected to increase by median values of 3.5 and 3.6°C, respectively, for the A1B scenario, with the largest increases expected during the summer months of June, July, and August. These temperature increases are larger than the global annual mean warming of 2.8°C predicted by the MMD models for the same period.[xiii]

The IPCC also used the MMD models to estimate precipitation changes for the 21st century under the A1B scenario. As shown in Table 2, the model simulations show a general decrease in rainfall across North Africa, with median decreases in average annual precipitation of 12 percent and 6 percent projected for the Mediterranean and Saharan regions, respectively. This general drying trend for North Africa is punctuated by seasonal variations in projected precipitation that differ by region. All of the MMD models show a clear decrease in future precipitation in the SEM region, with the largest decrease of 24 percent expected during the months of June, July, and August. This drying trend is likely driven by increased moisture divergence and a systematic northward shift of the storm tracks affecting winter precipitation in the region, as well as losses in soil moisture during the summer.[xiv] Results are less definitive for the SAH region; all of the MMD models predict a clear decrease in future precipitation of 18 percent for the period December to May, but the predicted decrease is much smaller for the months of June, July, and August. Furthermore, the models predict a slight increase in future precipitation of 6 percent for the months of September, October, and November in the SAH region. Some of these variations in predictions of future precipitation trends are likely related to uncertainty in the ability of climate models to successfully downscale precipitation over Africa.[xv]

Season	Temperature Change (°C)		Precipitation Change (%)	
	SEM	SAH	SEM	SAH
Annual	+3.5	+3.6	-12	-6
Dec/Jan/Feb	+2.6	+3.2	-6	-18
Mar/Apr/May	+3.2	+3.6	-16	-18
Jun/Jul/Aug	+4.1	+4.1	-24	-4
Sep/Oct/Nov	+3.3	+3.7	-12	+6

Table 2. **Median changes in temperature and precipitation predicted by a set of 21 IPCC MMD models under the A1B emissions scenario for the Southern Europe and Mediterranean (SEM) and Saharan Africa (SAH) regions.** Source: Intergovernmental Panel on Climate Change (IPCC), *Climate Change 2007: the Physical Science Basis*, eds. S. Solomon, D. Qin, M. Manning, M. Marquis, K. Averyt, M.M.B. Tignor, H.L. Jr. Miller, and Z. Chen (Cambridge: Cambridge University Press, 2007), http://www.ipcc.ch/ipccreports/ar4-wg1.htm.

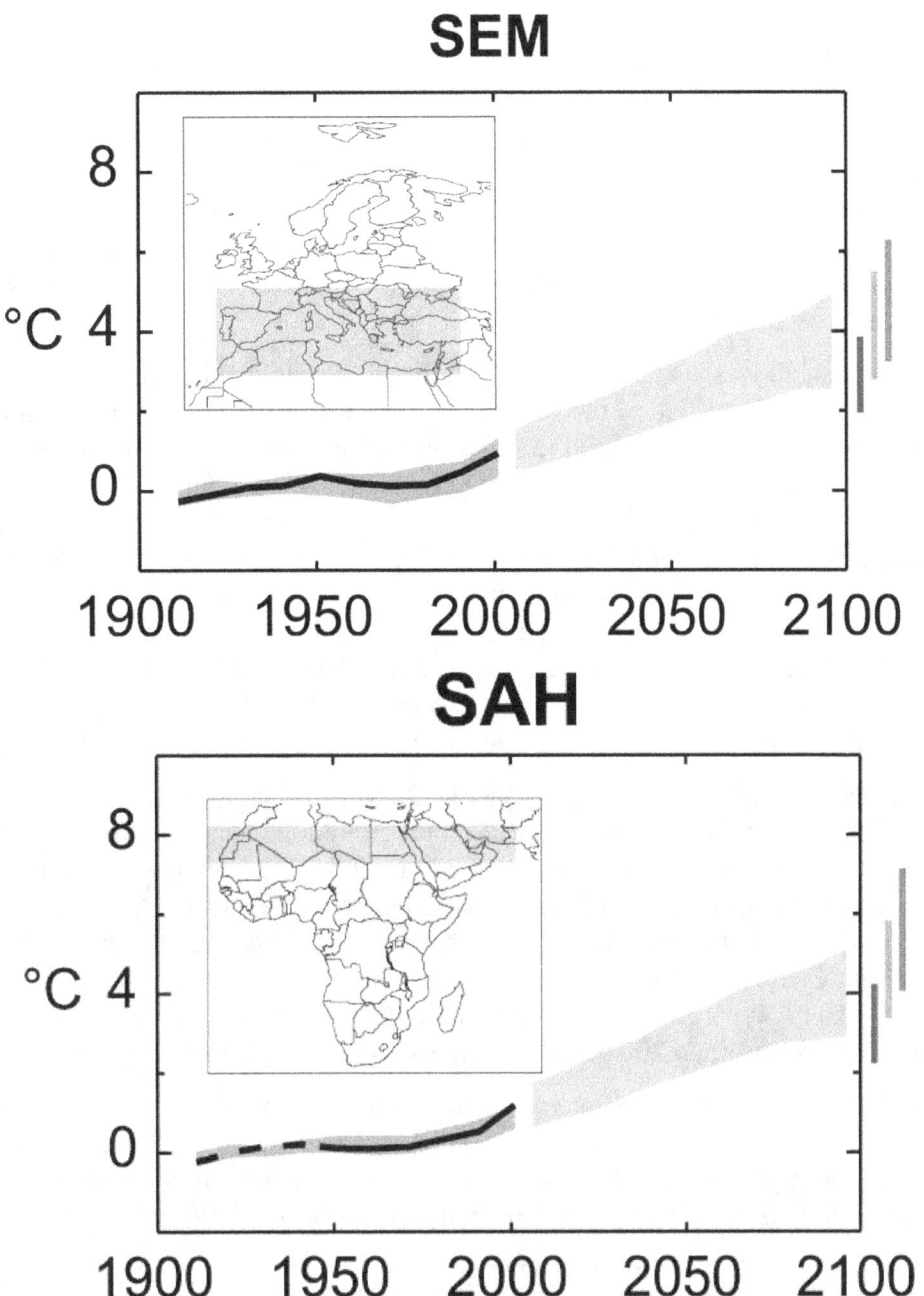

Figure 5. Temperature changes in °C predicted by the IPCC MMD models for the Southern Europe and Mediterranean (SEM) and Saharan Africa (SAH) regions. Temperature anomalies for the region with respect to 1901-1950 are shown for 1906-2005 (black line), as simulated by the MMD models using known forcings (red envelope), and as projected for 2001-2100 by the MMD models for the A1B scenario (orange envelope). The colored horizontal bars on the right side of the figure represent the range of projected changes for 2091-2100 for the B1 scenario (blue), the A1B scenario (orange), and the A2 scenario (red). The width of the shading and the bars represent the 5 to 95 percent range of the model results.[xvi]

The same trends in future temperature and precipitation changes predicted by the IPCC MMD models also have been simulated by several North African regional and country-specific studies. In an analysis of climate change in Africa through 2100, Hulme et al[xvii] used seven GCMs run under the A1, A2, B1, and B2 emissions scenarios to project temperature and precipitation changes for the "2020s" and "2050s" relative to 1961-1990. For the North Africa region, the models predicted temperature increases of 0.9 to 1°C and 1.8 to 2.2°C for the B1 and A2 scenarios, respectively, for the "2020s"; and increases of 1.2 to 1.5°C and 3.1 to 4.4°C for the B1 and A2 scenarios, respectively, for the "2050s." Precipitation projections were less definitive but involved a drying trend for the region, especially along the Mediterranean coast of North Africa, that is expected to become more pronounced with time.

The future drying trend was also observed in a study of anthropogenic climate change in the Mediterranean region for the end of the 21st century.[xviii] Results from a global variable-resolution climate model run under the B2 emissions scenario projected a decrease of 0.065 mm/day in annual average precipitation for the period 2070-2099 relative to 1960-1989. The authors predicted that evaporation in the Mediterranean basin would increase in winter and spring but decrease in summer and fall, despite projected increases in surface temperature, leading to the net decrease in regional precipitation. Giorgi[xix] found similar results in his development of a Regional Climate Change Index (RCCI) from the output of 20 GCMs run under the A1B, A2, and B1 emissions scenarios. Giorgi determined that the Mediterranean region was one of the two most vulnerable areas in the world to climate change due to the large decrease in mean precipitation predicted by the GCMs for the period 2080-2099.

Most recently, Paeth et al[xx] used the regional REMO GCM run under the A1B emissions scenario to estimate annual mean temperature and precipitation changes in tropical and North Africa through 2050. REMO is a synoptic scale model for Africa that has a spatial resolution of 5°. Paeth et al results confirm those of the IPCC and previous researchers, who predict a warming and drying trend in the region over the next century. Paeth et al estimate that by 2050, surface temperatures in North Africa will increase by approximately 1.5 to 2°C and precipitation will decrease by 10 to 30 percent across many of the desert areas of the region, with larger precipitation decreases of up to 200 percent along the coasts of Morocco, Algeria, and Tunisia.

These regional trends in temperature and precipitation are supported by a recent country-scale model analysis for Morocco.[xxi] Results from a statistical analysis of output from a series of GCMs suggest that mean annual surface temperatures in Morocco will increase by 0.6 to 1.1°C and annual precipitation will decrease by 4 percent for the period 2000-2020.

Projections of Future Changes in Sea Level
As the global ocean warms due to climate change, its volume will increase, and as a result, sea level will rise. The rate of sea level rise differs on a regional scale, principally due to local variations in the balance between the density and circulation of the oceans. Recent changes in Mediterranean Sea level along the North African coast are difficult to quantify due to lack of observational data. An analysis[xxii] of tide gauge data from the Permanent Service for Mean Sea Level (PSMSL) data set indicates that Mediterranean Sea level rose at a rate of approximately 1 mm per year during the 20th century. This value is based on data from stations situated along the southern European coast, however, and therefore may not be completely representative of recent changes in sea level along the coast of North Africa. Only a few tide gauge stations are located along the North African coast – near the Strait of Gibraltar and the Nile River Delta – and these stations

do not have a long enough record to reliably estimate changes in sea level during the 20th century.

The sparseness of tide gauge records is a global problem, and researchers have attempted to overcome this issue by combining the tide gauge observations with satellite altimeter data. Recent analysis of tide gauge records and data from the TOPEX/Poseidon satellite altimeter indicate that Mediterranean Sea level rose approximately 0.5 to 1 mm per year during the period 1950-2000.[xxiii] Thus, although definitive assessment of recent changes in Mediterranean Sea level is not possible, it seems likely that sea level rose along the coast of North Africa by approximately 0.5 to 1 mm per year for at least the second half of the 20th century. This increase is less than the global average sea level rise of 1.8 ± 0.5 mm per year for 1961 to 2003 and 1.7 ± 0.5 mm per year for the 20th century estimated by IPCC using PSMSL tide gauge records.[xxiv]

Using the ensemble mean of a subset of 16 models from the MMD, the IPCC has estimated future regional changes in sea level due to thermal expansion, including ocean density and circulation changes. The majority of future sea level rise is expected to be caused by the thermal expansion of oceans, and the rest will be due to melting ice sheets and glaciers, with minor contributions from land subsidence and changes in atmospheric pressure.[xxv] Results for the projected sea level rise due to thermal expansion for North Africa for the period 2080-2099 relative to 1980-1999 for the A1B emissions scenario are shown in Figure 6. These sea level projections are given in relation to the average global sea level increase of 13 to 32 cm due to thermal expansion that has been predicted by the IPCC MMD models for the same period under the A1B scenario. The global MMD climate models do not have sufficient spatial resolution to capture future changes in the Mediterranean Sea; therefore the only relevant projected changes in sea level indicated by Figure 6 for this report are for the Atlantic coast of Morocco. Figure 6 indicates that sea level along the Atlantic coast of Morocco will rise approximately 0 to 5 cm above the global average value at the end of the 21st century, for a net increase of 18 to 37 cm under the A1B scenario.

Since the global MMD climate models are not able to simulate future climate changes in the Mediterranean Sea region, downscaling and use of regional climate models are required. Only one climate model is currently available at the regional scale for accurately simulating future changes in the climate of the Mediterranean Sea region: the Sea-Atmosphere Mediterranean Model (SAAM). Developed by the Centre National de Recherches Météorologiques (CNRM) in France, SAAM is an ocean-atmosphere regional climate model with a horizontal resolution of 9 to 12 km.[xxvi]

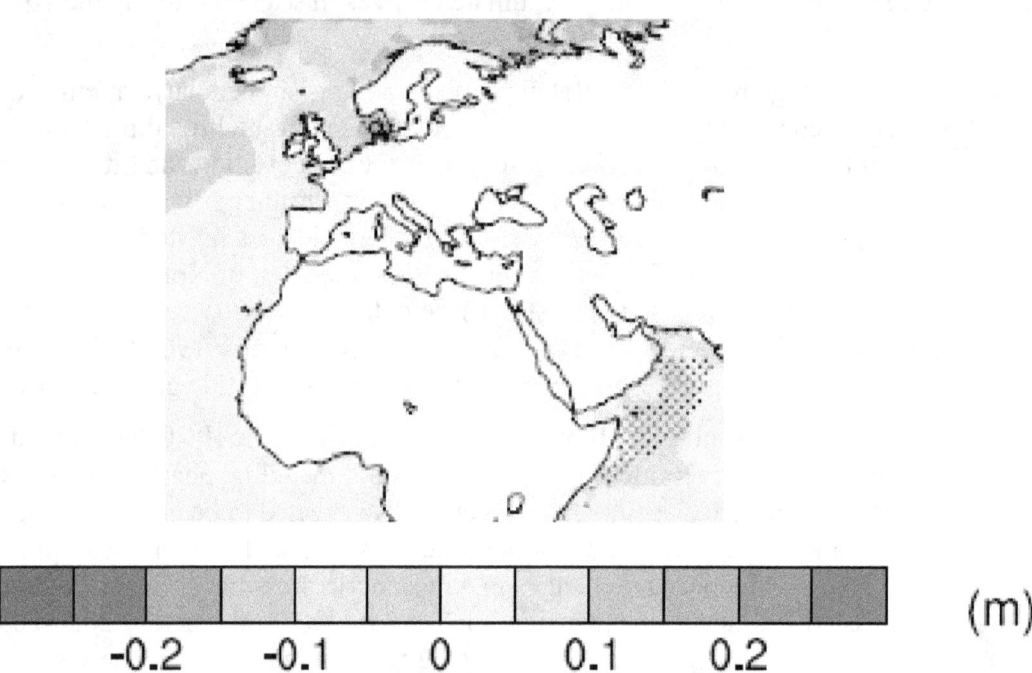

(m)

Figure 6. Projected changes in local sea level due to thermal expansion for the period 2080-2099 relative to the period 1980-1999. Sea level change was calculated as the difference between averages of projections for 2080-2099 and 1980-1999 from the ensemble mean of 16 GCMs run under the A1B emissions scenario. The local sea level changes shown are above/below the global average sea level rise of 13 to 32 cm due to thermal expansion that is expected by the end of the 21st century under the A1B emissions scenario. Changes in sea level associated with thermal expansion include changes in ocean density and circulation.[xxvii]

To predict future changes in Mediterranean Sea level rise, SAAM was run for the period 1960-2099 under the A2 emissions scenario.[xxviii] The model simulated changes in Mediterranean Sea level associated with thermal expansion of the oceans and did not include contributions from melting ice sheets and glaciers. Results show that projected sea level rise associated with changes in water density (on the order of 5 to 20 cm) is significantly larger than that associated with changes in circulation (on the order of 0 to 3 cm). The largest increase in sea level on the North African coast is expected to be approximately 15 to 20 cm (5 to 6.5 mm per year) along the Algerian coast. Mediterranean Sea level rise along the rest of the North African coast is expected to be approximately 5 cm (1.7 mm per year) or less. These regional Mediterranean Sea level projections are roughly comparable to the average global sea level increase of 3.8 ±01.3 mm per year due to thermal expansion that has been predicted by the IPCC MMD models for the period 2080-2100 under the A2 emissions scenario.[xxix] It is important to note that the regional Mediterranean Sea level projections are derived from a single model run; thus their results are significantly more uncertain than would analogous results from an ensemble of models, but they represent the most current estimates of sea level rise along the North African coast.

Older model projections provide a more pessimistic view of sea level increases in the North African region. A recent analysis[xxx] of the impacts of sea level rise on the Mediterranean coast of Morocco used model simulations from a 1996 modeling study[xxxi] that predicted sea level increases of 20 to 86 cm along the Moroccan coast by 2100 under the old IS92a emissions scenario. The IS92a scenario assumed a worldwide rate of greenhouse gas emissions that would cause a doubling of atmospheric CO_2 concentrations over pre-industrial levels by the end of the 21st century, which is too aggressive by about 50 years, according to recent estimates,[xxxii] thus making the accuracy of 1996 model study uncertain. In addition, an unspecified Tunisian model study, cited in the Initial Communication of Tunisia to the United Nations Framework Convention on Climate Change, projected a sea level increase of 38 to 55 cm along the Tunisian coast by 2100.[xxxiii]

Projections of Future Changes in the Frequency and Intensity of Extreme Climatic Events
Little research has been conducted on changes in the frequency and intensity of extreme climatic events in Africa, either in analysis of past observations or future model predictions. North Africa is not subject to tropical cyclones (hurricanes) or directly impacted by the El Niño – Southern Oscillation (ENSO) phenomenon. The most common extreme events in the region are droughts and floods. There is some evidence that the intensity and frequency of floods and droughts have been increasing in recent years, although we do not know to what extent these increases are due to the influence of climate change versus other anthropogenic effects, such as changes in land cover or irrigation practices. Catastrophic floods in Algeria and Morocco in 2001-2002 caused extensive destruction of property and loss of life.[xxxiv] The worst flooding event in Algerian history occurred in Algiers on November 10-14, 2001, when the equivalent of an entire month of rainfall fell in a matter of several hours. 751 people died and property losses were estimated at $300 million USD equivalent. In Morocco, 63 people died and hundreds of hectares of agricultural land were damaged from a devastating flood on November 20-27, 2002. In addition, climatic data from Morocco, Algeria, and Tunisia indicate that the frequency of droughts increased from approximately one per decade to 5-6 per decade during the course of the 20th century.[xxxv]

The IPCC[xxxvi] reports that in regions where mean drying is expected, such as North Africa in the 21st century, there will be a proportionally larger decrease in the number of rainy days expected to occur. This result suggests that although there will be less net rainfall in the region in the future, rain will be more intense on days that it does fall, indicating a compensation between intensity and frequency of precipitation that may lead to enhanced flooding.

In its First National Communication to the United Nations Framework Convention on Climate Change,[xxxvii] Morocco outlined several likely changes in extreme events, based on climate scenarios developed according to IPCC guidelines. In the 21st century, the frequency and intensity of frontal and convective thunderstorms in the regions north and west of the Atlas Mountains[5] are expected to increase, the frequency and intensity of droughts in the southern and eastern portions of the country are expected to increase, and the length of the rainy season is expected to decrease.

Sea level rise associated with climate change is expected to increase flooding along the coast from storm surges.[xxxviii] For example, a recent analysis of the potential increase in flooding along the eastern Mediterranean coast of Morocco[xxxix] predicts that for a sea level rise of 20 to 86 cm, 24 to 59 percent of the coastal area will be lost due to flooding in 2050-2100. Residential, recreational, and agricultural lands are expected to be most impacted.

---

[5] The Atlas Mountains extend across northern Africa from Morocco to Tunisia, separating the coastal areas from the Sahara Desert. Separate ranges in Morocco, running from southwest to northeast, include the Anti-Atlas, the High Atlas, and the Middle Atlas Ranges.

Due to lack of research, it is not possible to predict any definitive future trends in the frequency or intensity of extreme climatic events in North Africa. Droughts and floods are the most common extreme climatic events that occur in North Africa, and there is observational evidence that their frequency and intensity have increased in the past 10 to 20 years, especially in Morocco, Algeria, and Tunisia. Likely future changes across the region include periods of more intense rainfall and increases in coastal flooding from storm surges associated with sea level rise.

Impacts of Climate Change on Human-Natural Systems

According to Claudia Ringler (2009),[xl] a senior research fellow at the International Food Policy Research Institute, "While population, diet patterns, and urbanization currently have a greater impact on water security, by 2025, climate change will account for more of the threat." As pointed out by a number of studies reviewed in this report, the general drying trend predicted for North Africa may in turn translate to sizable impacts on the region's agriculture.

In 2005 the agricultural area of the three largest countries in North Africa—Algeria, Libya and Egypt—was estimated at about 60.3 million hectares (11.7 percent)[xli]. The size of agricultural area in Morocco and Tunisia is considerably larger—62.9 percent and 68.1 percent of land, respectively (2005 estimate). Almost all of Libya and Egypt's land is desert (94-96 percent).[xlii] The geography of land use and demographic patterns of both countries reveal that rainfall (water resources) is the dominant factor in Libya, and the Nile River the dominant factor (source of water) in Egypt.

One-quarter of northern Algeria is unproductive; only 3 percent of its land is cultivated, despite having 25 percent of the population involved in agriculture.[xliii] Similar to Algeria, only a small percentage of Egypt's land is arable, but, unlike Algeria, its land is largely productive and can be cropped two to three times annually. In Libya, almost all crops are grown for domestic consumption. Cereals are produced in the northwestern (Tripolitania) and eastern (Cyrenaica) regions of the country, while agriculture in the southwestern region (Fezzan) of the country is concentrated in the oases. Morocco is relatively self-sufficient in food production, but due to recent occurrences of drought, it has been forced to import grains during some years. In Tunisia, fertile land is usually limited to the north, where cereals, olives, fruits, grapes and vegetables are harvested. Currently, Tunisia's growing season provides the country an advantage by allowing it to profit from exporting fresh produce to Europe before European crops ripen.

Table 3 summarizes agricultural production, major exports and imports, and the total size of agriculture as part of each North African country's GDP.

Water resources and agriculture, being intertwined impacts, are discussed first, followed by a discussion of climate change impacts on migration, coastal areas, tourism and energy.

	Agricultural Production	Major Exports	Major Imports	Agricultural GDP as share of total GDP Percent, 2004
Algeria	Wheat Cow milk (fresh) Indigenous sheep meat	Dates, Oil of maize, Cocoa butter	Wheat, Dry whole cow milk, Dry skim cow milk	9.8
Egypt	Tomatoes Rice, paddy Buffalo milk	Cotton lint Rice, milled Oranges	Wheat Maize Cake of soybeans	15.1
Libya	Indigenous chicken meat Olives Indigenous sheep meat	Skins, dry salted (sheep) Crude organic materials, NES Skins with wool, sheep	Flour of wheat Oil of maize Wheat	---
Morocco	Wheat Cow milk, whole (fresh) Indigenous chicken meat	Tangerines, mandarins, clementines, satsumas Crude organic materials, NES Oranges	Wheat Maize Oil of soybeans	15.9
Tunisia	Olives Wheat Tomatoes	Oil of olive, virgin Dates Crude organic materials, NES	Wheat Cake of soybeans Maize	12.6

Table 3. 2006 Agricultural Outlook in North Africa. Information not available. *Source*: FAO (Food and Agricultural Organization of the United Nations), s.v. "Agricultural sector," FAO Country Profiles, 2006, http://www.fao.org/countryprofiles/.

Water Resources Stress

Observed trends pointing to a drier North Africa will have severe implications on the region's water availability, accessibility, and demand. Already, North Africa is experiencing high water stress. According to Ashton (2002),[xliv] even without climate change, North African countries will surpass their maximum economically usable land-based water resources before 2025. Population increases accompanied by growing water demands are causing significant water deficits in these arid countries. For the period 1990-2002, the average annual population growth in North Africa was the highest in the world at 2.9 percent.[xlv] At the same time, the Water Exploitation Index (i.e., total water extraction per year as a percentage of long-term freshwater resources) is high for most countries of the region: more than 50 percent for Tunisia, Algeria, and Morocco, and more than 90 percent for Egypt and Libya.[xlvi] Physical water scarcity[6] becomes apparent when withdrawals surpass 40 percent of the annually renewable resource.[xlvii]

Given an expected population growth in the region of approximately 50 million between 2025 and 2050,[xlviii] climate change-induced droughts are likely to place further stress on North Africa's freshwater resources. With an increase of 3°C, for example, an estimated 155-600 million North Africans will experience water stress.[xlix]

In preparation for Morocco's Initial Communication to the United Nations Framework Convention on Climate Change (UNFCCC), a partial study of vulnerability to climate change

---

[6] Physical water scarcity is water availability of less than 1,000 cubic meters per person per year.

impacts was conducted.[l] Based on this study, it is expected that, by 2020, climate change-related disruptions in rainfall will reduce dam capacity (i.e., concentrated rainfall and rapid sludge accumulation exacerbated by erosions); upset river flow rates; and decrease water levels, thereby effectively decreasing natural outlets for water tables and increasing salinity in coastline areas; and deteriorate water quality. The study estimated the result will be a 10-15 percent reduction of Morocco's water resources. This same reduction is estimated for Algeria.

The IPCC Assessment[li] further reports that projected temperature increases of 1-4°C and declines in precipitation of up to 10 percent are likely to result in annual reductions in runoff in Morocco's Ouergha watershed.[7] For example, holding precipitation levels constant, a 1°C rise in temperature could lead to a 10 percent reduction in runoff.[lii] Assuming annual runoff reductions in other watersheds, the aggregate decline is equivalent to the loss of one large dam per year in the region.

According to GTZ (2007),[liii] a private international enterprise that collaborated with various Tunisian ministries and authorities as well as several nongovernmental organizations to develop an adaptation strategy to climate change in the country's agricultural sector, Tunisia's water resources are estimated to decline 28 percent by 2030. Of particular risk will be the loss of groundwater reserves.

Yet a graver situation may be experienced in Egypt, where it is predicted that 74.8 percent of Egyptians will have less than adequate fresh water supply by 2030.[liv]

With climate change comes a looming potential for conflict over water resources. Over extraction of water together with the impacts of climate change may lead governments to divert major river dams, construct large dams, or tap underground aquifers that traverse beneath the territory of neighboring countries.[lv] Such may prove to be the case with the large fossil-water aquifers beneath the Sahara desert—namely, the Eastern Erg artesian aquifer, which extends from Algeria into Tunisia, and the Nubian Sandstone underlying Libya, Egypt, and Sudan.[lvi] There are already concerns that the latter is generating international conflict. Libya is tapping the Nubian aquifer on a massive scale as part of its "Great Manmade River" project.[lvii] This in turn is creating fear that pumping water beneath Libyan territory is draining groundwater reserves in Egypt and Sudan.

Similarly, with increased heat and aridity, conflicts over the Nile River may surface between Egypt and its neighboring countries. Egypt's dependence on the Nile River is appreciable; more than 95 percent of its water needs are satisfied from the Nile.[lviii] The Nile waters, which originate outside Egypt, traverse nine countries to the South: Sudan, Eritrea, Ethiopia, Uganda, Kenya, Tanzania, Rwanda, Burundi, and Zaire. Although the Nile River is governed by International Law (the Nile Agreement of 1959, signed by Egypt and Sudan), and despite relatively consistent cordial relations between Egypt and Sudan, disputes over water between the two countries have occurred. In 1995, in the wake of the assassination attempt on President Hosni Mubarak, accusations began flying back and forth between Cairo and Khartoum, and Sudan threatened to "cut off Egypt's water."[lix] The fact that Sudan lacks the engineering capacity to carry out such a threat did little to soften the frenzy in the Cairo press, and the current Arab League's Secretary-General, Amr Moussa, then former Egyptian Foreign Minister, warned Islamic leader Hassan al-Turabi not "to play with fire."[lx]

---

[7] The Ouergha River is a tributary of the Sebou River (the largest by volume in Morocco) in northern Morocco and the source of water for the Al Wahda Dam, the second-largest dam in Africa.

As less water becomes available, the demand for water grows higher and, with higher demand, the need to rely on costly water treatment and extraction methods increases. Wastewater treatment and re-use has become prevalent in Tunisia, Egypt and Morocco; the use of desalinated water has become prevalent in Egypt and Libya.[lxi] In addition, North Africa is increasingly investing in water infrastructure, accounting for more than 20 percent of public sector investment in Morocco and Tunisia, and 12 percent in Algeria.[lxii] In 2002, Ragab and Prudhomme[lxiii] reported that:

- By 2010, Algeria estimates it will need another 5.5 km^3 of water annually: 50 percent for irrigation and 50 percent for domestic and industrial uses. It plans to build 50 more dams and 10 diverting canals and will tap non-renewable fossil water beneath the Sahara.

- In 2000, Tunisia expected to use 90 percent of its surface water in the north and all of its groundwater. Part of the country's agenda is to build new large dams and develop a network of pipes and canals for water transportation between river basins. As a result of these efforts, Tunisia will be able to transport more than half of the water captured behind dams in the northern regions.

- Morocco plans to double the proportion of its river flow that is controlled by dams and extract more groundwater. It will construct 60 large dams, 100 km-long sink boreholes and construct 280 km of water-transportation structures.

- Libya intends to tap more underground water and transport it to the coastal aquifers, which have been excessively depleted due to overextraction. Libya's annual water withdrawal is larger than the volume of its renewable resources.

According to Egypt's Initial National Communication on Climate Change to the United Nations Framework Convention on Climate Change (UNFCCC), Egypt also is undergoing massive projects to divert some of the Nile waters to Northern Sinai and to the Toshka depression in the extreme southern part of the country.[lxiv] In addition, 70 percent of cultivated areas rely on low-efficiency surface irrigation systems, which result in high water losses, a decline in land productivity, saturation of the ground with water, and salinity problems.[lxv]

As the true costs of pollution and water scarcity become increasingly apparent, however, governments and policymakers are also becoming increasingly willing to address water problems.[lxvi] In 2007, Algeria, Egypt and Morocco spent 20 and 30 percent of their budgets on water. Bucknall's, *Making the Most of Scarcity: Accountability for Better Water Management in the Middle East and North Africa* (2007),[lxvii] assesses that by 2050, MENA governments will most likely begin to employ water for enterprises that generate the highest amount of money and employment and begin to import more hydro-intensive crops such as wheat. The report estimates that the cost of water-related environmental problems is between 0.5 and 2.5 percent of GDP for most countries.

Given that the volume of water available for food production has not been sufficient to satisfy increasing demand, Wichelns (2000)[lxviii] describes the role of virtual water (i.e., the volume of water embodied in food crops that are traded internationally) as a method by which North African countries have coped with water scarcity in the past. Such was the case during the boom of the 1970s, when higher incomes and increasing populations created a greater demand for food that could only be generated by increasing food imports.

Agriculture

Irrigation plays the most crucial role in the arid regions, which are characterized by low rainfall or evapotranspiration exceeding precipitation most of the year and high interannual rainfall variability. Accordingly, water scarcity related to climate change is expected to have negative consequences on North Africa's agriculture. Water and land resources in North Africa are primarily used for agriculture.[lxix] North Africa accounts for more than 41 percent (about 6 million hectares) of total irrigated lands in Africa.[lxx] Consequently, the Northern region represents more than half of the agricultural water withdrawal of the continent.[lxxi]

In North Africa, rising temperatures associated with climate change are expected to decrease the land areas suitable for agriculture, shorten the length of growing seasons, and reduce crop yields.[lxxii] The decrease in annual precipitation that is predicted for Northern Africa in the 21st century will exacerbate these effects, particularly in semiarid and arid regions that rely on irrigation for crop growth. Rising atmospheric CO_2 levels are expected to stimulate plant photosynthesis, however, which might result in higher crop yields and could offset some of the negative effects of higher temperatures and less rainfall in the region.

Only a few systematic modeling studies of the effects of climate change on agriculture in North Africa are available, possibly due to the relatively small areas of arable land in the region combined with the lack of suitable regional or downscaled climate model information. For example, only about 3 percent of the total land area of Egypt can be used for farming, almost all of it in the Nile River Delta and Valley.[lxxiii] GCMs cannot accurately resolve temperature and precipitation variations on such a small spatial scale, which makes evaluation of future agricultural changes difficult.

Already, however, the areas suitable for agriculture are heavily exploited.[lxxiv] Arable land availability, food price shocks, and population growth are current problems.[lxxv] In addition, policymakers in the region tend to adopt agricultural policies that favor perimeter irrigation, construction of dams, and the development of tourism sectors, which are water-intensive, over more traditional agricultural cultivation, which requires less water (e.g., wheat) and the development of fruits and vegetables for export.[lxxvi]

Based on observations, experiments, and model analyses made by the National Institute for Agricultural Research in Morocco, several potential changes to agricultural growing seasons in Morocco were outlined in Morocco's First National Communication to the United Nations Framework Convention on Climate Change.[lxxvii] Likely changes in the 21st century include a reduction in the growth period of regional crops, a reduction in the duration of crop cycles, and an increase in the risk of dry periods during the course of crop cycles.

Several researchers have used GCM climate simulations in conjunction with crop models in an attempt to quantify the effects of climate change on crop yields, particularly in Morocco and Egypt. Jones and Thornton (2003)[lxxviii] investigated the potential changes in maize production associated with climate change in Africa and Latin America, including Morocco, using the HadCM2 model. The authors focused on maize because it is an important crop for smallholder farmers in those regions. HadCM2 simulated changes in temperature and precipitation for the period 2040-2069 ("2055"), and these climate projections were used by the CERES-Maize crop model to simulate the growth, development, and yield of maize crops. Model results predicted a substantial increase in maize crop yield in Morocco in 2055 due to the effects of climate change, from a baseline value of 317 kg/ha to a value of 550 kg/ha, which represents an approximately

175 percent increase. These model simulations suggest that the positive effects of increased atmospheric CO_2 concentrations may be the dominant influence on future maize crop yield in Morocco. Maize is a member of a group of plant species that have a relatively high optimum temperature range for photosynthesis,[lxxix] which makes maize tolerant of high air temperatures and may explain why higher crop yields were projected for a hotter future climate.

Egypt, where agriculture is not feasible without irrigation (99.8 percent of its cropland is irrigated), depends on other countries for over 90 percent of its renewable water resources.[lxxx] Almost all of Egypt, 96 percent of its land, is covered by desert, and 97 percent of the population is concentrated on only 4 percent of irrigated land.[lxxxi]

In 2006, Egypt's cultivated area was estimated at approximately 8 million acres or about 3.5 percent of the country's total land area.[lxxxii] Meanwhile, 30 percent of Egypt's labor force works in the agricultural sector; contributing 14.8 percent of the nation's GDP (2006 estimate).[lxxxiii] Arable land in Egypt is restricted to the Nile valley from Aswan to Cairo and the Nile Delta north of Cairo.[lxxxiv] The Nile Delta, a region characterized by high production, high irrigation, and smallholder agriculture, is also a region with severe urban water and land-use challenges and projections of high population increase (see Natural Disasters discussion below).[lxxxv]

In recent years, the country has undergone several major agricultural developments, including expanding production of two of its major exporting crops as shown in Table 4: rice (40 percent of agricultural exports) and cotton (20 percent). However, these crops are also two of the most hydro-intensive crops. A changing climate, together with the expansion of cultivated areas in Egypt given the country's policy to add more agricultural lands, implies additional stress on the country's water resources and all the negative ramifications on agriculture and its economy.

Produce	Unit	Productivity Feddan (Acres)			Realized Percentage %
		2001/02 Base year	2006/07 Targeted	2006/07 Achieved	
Wheat	Ardeb	18.8 (19.5)	20.0 (20.8)	19.25 (19.98)	96.3
Maize	Ardeb	24.75 (25.69)	27.5 (28.5)	25.5 (26.5)	92.7
Rice	Ton	3.9 (4.0)	4.2 (4.4)	4.2 (4.4)	100.0
Beans	Ardeb	8.6 (8.9)	10.5 (10.9)	9.2 (9.5)	87.6
Cotton	Cantar	7.2 (7.5)	8.0 (8.3)	7.6 (7.9)	95.0
Sugar Cane	Ton	50.0 (51.9)	51.5 (53.5)	51.4 (53.3)	99.8

Table 4. "Development in Agricultural Productivity during the 5-year-plan (2002-2007)"
Source: Egypt State Information Service: Your Gateway to Egypt, s.v. "Agriculture," http://www.sis.gov.eg/En/Economy/Sectors/Agriculture/050301000000000001.htm (accessed April 6, 2009).

In an early analysis of the effects of climate change on Egyptian agriculture, El-Shaer et al (1997)[lxxxvi] used the GFDL (Geophysical Fluid Dynamics Laboratory), UKMO (United Kingdom Meteorology Office), and GISS (Goddard Institute for Space Studies) GCMs, under unspecified conditions of $2 \times CO_2$, in conjunction with CERES crop models to simulate changes in wheat and maize yields and growing season length. Analysis focused on four important agricultural regions in Egypt: the northern Nile Delta, the Mid-Delta, the northern Mediterranean coast, and Upper (southern) Egypt. The models projected a decrease in wheat yield of 10 to 50 percent for all regions except the northern coast, where an increase of approximately 80 percent was predicted.

Wheat grows under natural precipitation in the northern coastal zone, in contrast to the other regions, where crops are irrigated, and the increases in wheat yield simulated by the models for the northern coast were tied to increases in precipitation projected for the region. The models also predicted a decrease in maize yields for all regions, with the largest losses of approximately 60 percent in Upper Egypt. These results for maize yields in Egypt are in contrast to those of Jones and Thornton[lxxxvii] for Morocco, which suggests that, in Egypt, the negative effects of future temperature increases may outweigh the positive effects of future atmospheric CO_2 concentration increases. The model results of El-Shaer et al (1997)[lxxxviii] also showed a decrease in growing season length in all regions for wheat and maize; the largest decreases were approximately 20 to 25 days in the northern Delta and northern coastal regions for wheat and in the Mid-Delta and Upper Egypt regions for maize. The authors found that simulation of adaptation strategies, such as shifting planting dates and changing crop varieties, made no significant impact on projected crop yields or season lengths.

In a follow-up study, Yates and Strzepek (1998)[lxxxix] examined the potential effects of climate change on crop yields in the Nile Basin in 2060 using the GFDL, UKMO, and GISS GCMs. Temperature and precipitation projections from the GCMs were used in conjunction with an unspecified crop model to estimate the changes in yield for wheat, rice, grains, protein feed, other food, non-food, and fruit yields in 2060 relative to 1990. All models indicated a decrease in crop yield, with the largest losses, up to 20 to 50 percent, in yields of wheat and grains. Climate change in the GCMs was represented by $2 \times CO_2$ equivalent to 600 to 640 ppm, which is a high projection relative to current estimates.[xc] Therefore, these model results should be interpreted with caution. The authors found that simulated adaptation measures, including increased fertilizer application and development of new crop varieties, substantially mitigated the projected decreases in crop yields expected in the mid-21st century, which is in contrast to the earlier findings of El-Shaer et al[xci] for the same region.

Most recently, Hegazy et al (2008)[xcii] investigated the influence of increased air temperatures associated with future climate change on the spatial and temporal distribution of four crops in Egypt through 2100. Analysis focused on cotton, wheat, rice, and maize because they are some of the most economically important crops currently grown in Egypt. Air temperature patterns in Egypt for 2025, 2050, 2075, and 2100 were simulated using a database of information compiled from multiple GCMs run under the B2 emissions scenario. Subsequently, these temperature projections were used to create seasonal and spatial crop distribution maps in a geographic information system (GIS) program. Crop distribution simulations were based on the optimum air temperatures for maximum growth of cotton, wheat, rice, and maize throughout the agricultural season, which are 23.0°C, 16.8°C, 25.8°C, and 26.0°C, respectively. Crop distributions were projected for three areas of the country: Upper (southern), Middle, and Lower (northern) Egypt. Currently, all four crops are grown in Lower Egypt; wheat, rice, and maize are grown in Middle Egypt; and only rice and maize are grown in Upper Egypt. Results from the model study indicate that higher temperatures in Egypt associated with climate change in the 21st century will likely necessitate a shift in crop sowing dates to earlier in the season, to prevent crop losses due to excessively warm growing conditions. Compared to the reference sowing year of 2005, the model study predicts future sowing dates will shift one to eight weeks earlier, depending on the crop type, region, and year. The exception is rice sowing in Upper Egypt, which is expected to shift one week later in 2025 and remain unchanged in 2050 and 2075. Wheat is the most heat-sensitive of the four analyzed crops, and the models estimate that by 2100, air temperatures will be high enough that growing wheat in Egypt will be impossible. During the 21st century, increases in air temperature are also predicted to cause a decrease in the land area suitable for growing wheat, from approximately 106,000 ha of land in Lower Egypt in 2005 to 6,500 ha in 2075. Cotton, rice, and maize are expected to be more tolerant of potential higher temperatures in Egypt and are not predicted to lose such a dramatic amount of viable growing area, assuming appropriate adaption measures are taken, such as shifting to earlier sowing times.

Using the MAGICC/SCENGEN and GCMs climate change scenarios, Eid et al (2007)[xciii] measured the economic impacts of climate change on farm net revenue in Egypt by 2050. The study was based on the Ricardian approach, which carries the advantage of accounting for a variety of adaptations (e.g., adoption of new crops and farming systems) that farmers apply in the face of a changing economy and environment.[xciv] The study found that if no adaptation measures are taken, temperature increases of 1.5°C (as predicted by the MAGICC/SCENGEN scenario) and 3.6°C (GCM scenario) will considerably reduce the farm net revenue per hectare. Accordingly, the study predicts that by 2050, climate change could reduce agricultural yields of several Egyptian crops—from an 11 percent reduction in rice to 28 percent for soybeans when compared to production under current climate conditions.

However, the same study suggests that reductions of farm net revenue could be less severe if farmers opt to use heavy machinery on farms, and revenue could even increase if farmers use irrigation.[xcv] Still, the study only takes into account changes in net farm revenue due to temperature changes and does not consider the effects of warming on water resources. Hence, predicted increases in farm net revenue due to irrigation could be offset by the impacts of climate change on the country's water resources.

Plans to increase agricultural productivity through intensive irrigation may not only imply further water scarcity, but also soil salination and associated desertification. Salination is a soil threat (inland) arising from a high evaporation rate under increasing temperature and reduced rainfall.[xcvi] Salination also represents a water threat due to over abstraction by a growing population, which can lead to sea water intrusion in coastal areas and the transformation of fresh groundwater into brackish water. The use of fossil groundwater to irrigate agricultural land can lead to salination, and is not a sustainable agricultural practice.[xcvii]

Soil salination already has been observed in several Mediterranean regions, including parts of Spain, Italy, and Greece. This trend, however, is particularly pronounced in countries south of the Mediterranean, such as in large parts of Algeria, Libya, and Egypt and a few regions in Morocco and Tunisia. For example, in Tunisia water quality is often a concern, as more than 30 percent of available water contains more than 3g per liter of salt.[xcviii]

In North Africa, salination is widespread and at risk of increasing. The impact on North African economy, compared to the impact on the region's European neighbors, is greater. Countries such as Morocco and Egypt have a considerably larger portion of their population employed in agriculture (44 percent and 28 percent respectively).[xcix] In addition, the contribution of agriculture to GDP in 2007 is also higher in North Africa—at 16, 14 and 11 percent in Morocco, Egypt and Tunisia, compared to 3 and 2 percent in Spain and Italy, respectively.[c]

Present Climate Change Coping Practices in Agriculture of Egyptian Farmers

A survey conducted by Eid et al (2007)[ci] of 900 Egyptian households from 20 governorates, revealed that Egyptian farmers have noticed a change in temperature and rainfall patterns—either from their own experience and/or with the help from agricultural extension teams. Overall, 85 percent of the households noticed a change in temperature in the form of heat waves during the summer and an increase in winter minimum temperatures. In addition, 65 percent of the households cited shortages in the amount of rainfall each season.

In response to the observed changes in temperature, many farmers adapted by increasing the frequency of irrigation, increasing the quantity, or avoiding irrigating in the afternoon when the temperature is at its highest. Some farmers reported changing their crop sowing dates to evade expected high temperatures, while others started using heat-tolerant varieties. Other reported changes included increased management of pesticide and fertilizer applications, planting trees as fences around the farm, use of intercropping between crop plants of varying heights, and fruit mulching for vegetables.

Regarding decreased rainfall, farmers recognized the use of high-water-efficient varieties and/or early maturing varieties as effective ways to cope with rainfall shortage, while others mentioned underground or drainage water for irrigation and improved drainage as alternative methods for coping with observed changes in rainfall.

Additional anthropogenic disturbances such as deforestation, overgrazing in rangelands, non-sustainable irrigation practices, and extractive farming practices, which produce fertility reductions and depletion of carbon stored in the soil, may be contributing to desertification in North Africa.[cii] According to Zafer Adeel, Director of the United Nations University's (UNU's) Canadian-based International Network on Water, Environment and Health and co-chair of the team that developed a global assessment of desertification as part of the 2005 Millennium Ecosystem Assessment, efforts in the past to forestall desertification have been consistently under-funded, and policies in the region that promote agricultural intensification in dry areas and the settlement of nomadic populations are exacerbating soil salination.[ciii]

According to Arnell (2004),[civ] the IPCC SRES scenarios indicate that by 2050, average annual runoff in North Africa will decrease considerably, effectively accelerating soil degradation in the region. If the trend toward soil salination persists, vegetation in the region will be compromised and desertification, where land is completely lost for agricultural use, is likely to expand. Desertification, would in turn imply increased emissions of carbon dioxide (as vegetation is lost) and other greenhouse gases, reductions in agronomic productivity, contamination of water resources, and reductions in biodiversity.[cv]

Since the beginning of the 20th century, Morocco has experienced droughts with a mean cyclic temporal frequency of 11 years.[cvi] However, the Moroccan Meteorological Office has observed a rise in the frequency, intensity, and duration of drought in the past three decades—an occurrence that is particularly pronounced in the spring.

In 2007, a drought caused a significant decline in Morocco's grain crop production: from 9.3 million tons in 2006 to 2.0 million tons in 2007.[cvii] According to the IPCC Chairman, Rajendra

Pachauri, the rise in the frequency and persistence of droughts due to rising world temperatures will likely increase the dependence on large-scale and costly food imports of countries such as Morocco.

Tunisia is also experiencing persistent droughts. Rain-fed agriculture represents 90 percent of the country's agricultural area, exposing this sector to climate variability.[cviii] Of particular importance to Tunisia's economy are cereals, which are primarily (97 percent) cultivated under rain-fed conditions. In the late 1990s, water reserves did not satisfy the water needs of both Tunisia and Morocco, which resulted in several irrigation-dependent agricultural systems to cease production.

Migration

Experts agree that the poorest countries will be the most affected by climate change.[cix] In particular, climate change poses a unique challenge for Sub-Saharan Africa, which comprises the poorest countries in the world. On one side, much of Sub-Saharan Africa is experiencing *economic* water scarcity, a condition in which a country cannot afford to make use of its water resources.[cx] On the other side, desertification is shifting the desert's limit further south in the Sahelian zone, particularly in Burkina Faso and Mali, as indicated by de Wit and Stankiewicz (2006).[cxi] As the effects of climate change further exacerbate these trends, North Africa undoubtedly will experience greater migration from Sub-Saharan Africa in the next few decades. However, there is little literature that deals with climate as a motivating factor for migration, and what references there are consider environment-migration relations at a very general level.[cxii]

Forced migration is a lingering problem in several countries of Africa, predominantly in Sub-Saharan Africa. Migrants residing in refugee camps or settlements for more than five years (known as protracted refugees) are common.[cxiii] In general, different estimates indicate that in any given year, between 65,000 and 120,000 Sub-Saharan Africans enter into Morocco, Tunisia, Algeria, Libya, or Mauritania, of which 70 to 80 percent are thought to migrate through Libya, and 20 to 30 percent through Algeria and Morocco.[cxiv] Specifically, more than 100,000 Sub-Saharan Africans currently live in Algeria (primarily Sahwari refugees), 1-1.5 million Sub-Saharan immigrants settle in Libya, and between 2.2 and 4 million in Egypt (primarily Sudanese).[cxv] In Tunisia and Morocco, the number of Sub-Saharan communities is smaller, but immigration from Sub-Saharan Africa into these two countries is also growing—on a magnitude of several tens of thousands of immigrants annually.[cxvi]

In addition, North African countries are in a special position with respect to migration in that they are both migration destinations and transit areas for refugees from Sub-Saharan countries and Asia trying to reach Europe (see http://www.migrationinformation.org/pdf/).[cxvii] In the context of North Africa as a transit region, several tens of thousands of Sub-Saharan Africans attempt to cross the Mediterranean every year.[cxviii] A growing number of Egyptians cross the Mediterranean to Italy by way of Libya. In addition, since Libya's pan-African policies connecting East African migration systems with the Euro-Mediterranean migration system, migrant workers and refugees from Sudan, Somalia, Eritrea, and Ethiopia, who used to settle in Cairo, now increasingly migrate to Libya by way of Sudan, Chad, or Egypt. There is also an increasing trend of Chinese, Indian, Pakistani, and Bangladeshi migrants who are flying to Morocco using Saharan routes through Niger and Algeria.

Besides cross-border migration, another issue exists in the increased urbanization of North African countries. There are two aspects. First, North African countries experience vast internal migration. The most common scenario involves migration from rural areas to the city. With an expected population increase of 50 million in North Africa (including, in this study, the five countries covered here plus Sudan) between 2025 and 2050, internal migration from rural areas to cities can be expected to increase in the first half of the 21st century.[cxix] Predicted sea-level rise in the coasts and reduced precipitation in many rural agricultural parts of North Africa will likely cause an exodus from rural areas to cities. Second, migrants who consider North Africa as their final destination or those who fail to enter Europe sometimes remain in North Africa. Presently, North African cities, such as Nouakchott, Rabat, Oran, Algiers, Tunis, Tripoli, Benghazi, and Cairo, are homes to large communities of Sub-Saharan migrants.[cxx]

There are two types of migration patterns identified by the IPCC assessment that may emerge as a result of climate change: (1) "repetitive migrants," representing ongoing adaptation to climate change, and (2) "short-term shock migrants," who respond to a particular climate event.[cxxi] Migration due to climate change may have the following implications:

- ***Greater Demands on Infrastructure Along the Mediterranean Coasts***.[cxxii]

- ***Violence Against Migrants***. Social unrest, including attacks against migrants stemming from ethnic and racial clashes as well as human trafficking, is already a reality for some North African countries.[cxxiii]

- ***Radical Fundamentalist Religious Movements***. Presently, for example, al-Qa'ida has bases in North Africa, including Algeria, and kidnappings of foreigners have occurred as recent as 2009.[cxxiv] The number of terrorist events perpetrated by radical fundamentalist groups may rise as a result of socio-economic factors related to climate change.[cxxv]

One of the few studies examining climate as a driving force for migration, Meze-Hausken (2000),[cxxvi] investigated the empirical consequences of climate perturbations on human adaptation and migration in dryland Africa. The study's objective was to investigate whether experience from past drought behavior during the previous three decades served as an analogy for impacts under future climate change. Vulnerability to climate events was assessed by analyzing migrants' behavior and living conditions before and after the onset of past droughts; specifically, 104 subsistence farmers in mixed agricultural systems in Tigray, Ethiopia were visited. The farmers were former drought migrants living in arid regions with high interannual rainfall variability. The results showed that although differentiation in farming yield is little during times of drought (i.e., at first, all farmers cultivate in a similar ecological setting, with little irrigation and similar technology), a combination of different socio-economic and environmental indicators such as animal holdings, non-agricultural income, and remittances determined how soon problems of shortage began after a drought. It was found, however, that after a certain number of months among critical food- and water-deficiency, the primary difference in vulnerability between households diminished. After a threshold is surpassed, farmers have no options to cope with the crisis, leaving farmers equally affected despite their socio-economic point of departure; many are forced to migrate. Failures of the response mechanisms of households and the relief mechanism of the state were the main drivers behind migration.

Natural Disasters Along Coastal Zones

According to Nicholls et al (1999) and Nicholls and Tol (2006), based on HadCM2 and HadCM3 models, the southern Mediterranean (Turkey-Algeria) is one of the regions with the largest risk of increased flooding in absolute terms for the 21st century.[cxxvii] The other regions of risk are West Africa (includes Morocco), East Africa, South Asia and Southeast Asia. Coastal cities will be especially vulnerable due to the concentration of poor populations in potentially hazardous areas, such as Alexandria—the second largest city in Egypt.[cxxviii]

Specifically, studies on the vulnerability of several sectors of Alexandria suggest that with a 30cm sea level rise by 2025, Alexandria will incur land and property losses of tens of billions of dollars, more than half a million inhabitants may be displaced, and 70,000 jobs will be lost.[cxxix]

Similarly, examining global model predictions of sea level rise by 2050 for 84 developing countries, a study by Dasgupta et al (2007)[cxxx] found that although sea level rise impacts on the land areas of the Middle East and North Africa (MENA) region are lower than the average of developing countries (0.25 percent vs. 0.31 percent with a 1m sea level rise), impacts on MENA are comparatively higher when measured in social, economic and ecological terms. Low-lying coastal areas in Tunisia, Libya, and particularly Egypt, were identified as having the greatest risk in North Africa.

According to Dasgupta et al, given a 1m sea level rise, an estimated 10 percent of Egypt's population would be impacted, with most of the impact being felt in the Nile Delta.[cxxxi] Thus, the impact of sea level rise will have subsequent effects on Egypt's agriculture and GDP (Figures 15).

To a lesser extent, approximately 5, 3, and 2 percent of the population of Tunisia, Libya, and Morocco will be impacted by a 1m sea level rise, respectively.[cxxxii] Based on Dasgupta et al's 2007 study, an assessment led by the World Bank (2007)[cxxxiii] estimated that for an increase in temperature of 1-3°C, between 6 and 25 million people will be exposed to coastal flooding in North Africa's urban areas.

However, given the lack of tide gauge observational data in the region, the wide range of future estimates in sea level, and the paucity of regional climate model projections for the Mediterranean Sea, a definitive estimate of sea level rise along the coastline of North Africa in the next 20 years is not possible.

Tourism

Tourism development is a major priority for North African governments, and the sector is expected to continue to increase in the next decade for most countries of the region (Table 5).[cxxxiv] In Tunisia and Morocco, coastal tourism is an important source of income, as is reef ecotourism in Egypt.[cxxxv] Most of the development, however, has been carried out without much regard for environmental protection, sewage treatment, water provision, and energy inputs.[cxxxvi] Particularly in light of the region's water scarcity, tourism development is of concern.

In Tunisia, for example, where "mass-beach" tourism is advertised, tourist zones per person consume approximately 8-10 percent times per day the quantity of water of the rest of the country.[cxxxvii] In Morocco, tourism development has resulted in large stretches of its coastlines turning into concrete.[cxxxviii] Declining water availability, sea level rise, and increasing temperatures could reduce not only the attractiveness, but also the livability of these ecosystems.

	2009 (%)	2019 (%)
Egypt	15.0	14.6
Libya	8.6	10.1
Tunisia	16.7	16.3
Algeria	4.3	5.0
Morocco	16.2	16.7

Table 5. Outlook of the contribution of the tourism sector to Gross Domestic Product in North Africa. Source: World Travel and Tourism Council, s.v.v. "Egypt," "Libya," "Tunisia," "Algeria," and "Morocco" http://www.wttc.org/eng/Tourism_Research/ Tourism_Economic_Research/Country_Reports/ (accessed April 26, 2009).

Energy

Because of their high natural gas and oil revenues, Libya and Algeria's economies (and to a lesser extent that of Egypt's), are likely to be less impacted by climate change in the short term (see Adaptive Capacity section for Libya). Morocco's dependence on agricultural production has slowed down its economic growth—a trend which may be exacerbated by climate change by the middle of the 21st century, unless significant policy changes are made.[cxxxix] For example, in 2005, Morocco's government spent more than twice the amount it had planned on food subsidies. By contrast, Tunisia has a more diversified economy. Hence, despite higher energy costs, Tunisia's economy has grown in recent years.

Algeria, Egypt, Libya, and Nigeria are the largest natural gas producers in Africa and the largest African consumers of natural gas.[cxl] The latter can be explained by the limited infrastructure there is on the continent for intraregional trade of natural gas. According to the Energy Information Administration (EIA)'s *International Energy Outlook 2008*, from 2005 to 2030, natural gas consumption will grow by 3.5 percent in Africa.

Libya has the largest proven reserves in Africa (41.5 billion barrels in 2007), followed by Nigeria (36.2) and Algeria (12.3)—all three are members of the Organization of Petroleum Exporting Countries (OPEC).[cxli] Libya's economic growth is contingent on the hydrocarbon industry, both because of the revenues from exports, and because the country relies on oil and natural gas to satisfy national demand. The World Bank estimates that in 2004 Libya's hydrocarbon exports contributed over 95 percent of total merchandise exports and revenues from the oil and natural gas sectors.[cxlii] In addition, revenues from these sectors equated to more than half of the country's GDP.

Similarly, in Algeria, oil and natural gas exports, which represented 98 percent of all exports in 2006, are the main driving force behind the country's significant economic growth in recent years.[cxliii] In 2006, Algeria's real GDP growth rate was 4 percent. Producing 2.8 trillion cubic feet of natural gas in 2004, Algeria ranks as the eighth largest natural gas producer in the world and the second largest producer among OPEC member countries after Iran. Like Libya, Algeria depends on fossil fuels to satisfy domestic demand. In 2004, natural gas comprised 62 percent of total energy consumption. The rest was exported primarily to Europe, although some was exported to the United States.

In the 1990s Brauch[cxliv] predicted that Libya and Algeria's fossil reserves would be depleted considerably in the 21st century; in particular, their oil reserves were estimated to be exhausted by 2014 and 2037, respectively. According to *Wood Mackenzie* reports,[cxlv] however, Libya remains highly unexplored and only 25 percent of the country is covered by exploration agreements with oil companies. In addition, since the 1990s, Libya's natural gas production has augmented considerably over the past few years (39 billion cubic feet in 2005). The country plans to free up more oil for export, particularly to Europe.[cxlvi] Although Algeria has produced oil since 1965, it, too, is considered to be underexplored.[cxlvii]

However, Algeria and Libya's economies are highly dependent on Europe. For example, in 2006, the vast majority of Libyan oil exports were sold to European countries, such as Italy (495,000 barrels/day), Germany (253,000), Spain (113,000), and France (87,000).[cxlviii] The United States, too, has increased its oil imports from Libya since lifting sanctions against Libya in 2004. In 2006, the United States imported an average of 85,500 barrels/day of total Libyan exports, compared to 56,000 barrels/day in 2005.

Research by Hans Günter Brauch indicates industrialized countries need to reduce their carbon dioxide emissions by 80 percent by 2050 in order to limit climate change.[cxlix] This might imply a shift in Europe's energy demand and a shift from fossil fuels to non-fossil energy sources. Accordingly, it will become critical that North African countries develop their comparative advantage in high solar energy or other industries for possible export in the future. As of yet, however, there has not been any indication that Europe will considerably reduce its natural gas and oil imports from North Africa.

In terms of emissions from North Africa, the 2007 Observatoir Méditerranean de l'Energie (OME) trend scenario, based on estimates provided by Mediterranean countries and their major energy companies, reveal that primary energy demand in the Mediterranean Basin will be 1.5 times higher in 2025 than in 2006.[cl] Emissions from energy use in southern and eastern Mediterranean countries in 2004 were estimated at 663 million metric tons of CO_2—an increase of 58 percent from 1990. This growth rate surpasses the global rate by 20 points. Throughout the Mediterranean, emissions from electricity and heating emit more CO_2 emissions from energy use than any other sector (e.g., transportation, construction, and other fuel combustion), especially in the southern and eastern Mediterranean countries. The 21 Mediterranean countries studied by OME only contributed 7.4 percent of total global emissions of CO_2 related to energy use between 1850 and 2005. Yet OME predicts that if investment and development decisions made on energy over the past 30 years remain the same, and in light of the demographic and economic trends of southern and eastern Mediterranean countries, an appreciable high growth in emissions could be expected south of the Mediterranean.

Adaptive Capacity

The impacts of climate change will be felt differentially, depending upon how well a society can cope with or adapt to climate change, that is, its adaptive capacity. Adaptive capacity is defined by the IPCC as "The ability of a system to adjust to climate change (including climate variability and extremes) to moderate potential damages, to take advantage of opportunities, or to cope with the consequences."[cli] Although the specific determinants (or "drivers") of adaptive capacity are debated among researchers, there is good agreement that economic, human, and environmental resources are essential elements. Some components of this adaptive capacity are near term, such as the ability to deliver aid swiftly to those affected by, e.g., flooding or droughts. Other components include a high enough level of education to enable workers to change the means of earning their livelihoods, sufficient unmanaged land that can be brought into food production, and institutions that provide knowledge and assistance in times of change. For instance, Yohe and Tol[clii] identified eight qualitative "determinants of adaptive capacity," many of which are societal in character, although the scientists draw on an economic vocabulary and framing:

1. The range of available technological options for adaptation.

2. The availability of resources and their distribution across the population.

3. The structure of critical institutions, the derivative allocation of decisionmaking authority, and the decision criteria that would be employed.

4. The stock of human capital, including education and personal security.

5. The stock of social capital, including the definition of property rights.

6. The system's access to risk-spreading processes.

7. The ability of decisionmakers to manage information, the processes by which these decisionmakers determine which information is credible, and the credibility of the decisionmakers themselves.

8. The public's perceived attribution of the source of stress and the significance of exposure to its local manifestations.

North African Adaptive Capacity in a Global Context
Researchers have only recently taken on the challenge of assessing adaptive capacity in a comparative, quantitative framework. A comprehensive global comparative study[cliii] of resilience to climate change (including adaptive capacity) was conducted using the Vulnerability-Resilience Indicators Model (VRIM—see box below).

Adaptive capacity, as assessed in this study, consists of seven variables (in three sectors), chosen to represent societal characteristics important to a country's ability to cope with and adapt to climate change:

Human and Civic Resources

- *Dependency Ratio:* proxy for social and economic resources available for adaptation after meeting basic needs.

- *Literacy:* proxy for human capital generally, especially the ability to adapt by changing employment.

Economic Capacity

- *GDP (market) Per Capita:* proxy for economic well-being in general, especially access to markets, technology, and other resources useful for adaptation.

- *Income Equity:* proxy for the potential of all people in a country or state to participate in the economic benefits available.

Environmental Capacity

- *Percent of Land that is Unmanaged:* proxy for potential for economic use or increased crop productivity and for ecosystem health (e.g., ability of plants and animals to migrate under climate change).

- *Sulfur Dioxide Per Unit Land Area:* proxy for air quality and, through sulfur deposition, other stresses on ecosystems.

- *Population Density:* proxy for population pressures on ecosystems (e.g., adequate food production for a given population).

Methodological Description of the Vulnerability-Resilience Indicator Model (VRIM)

The VRIM is a hierarchical model with four levels. The resilience index (level 1) is derived from two indicators (level 2): sensitivity (how systems could be negatively affected by climate change) and adaptive capacity (the capability of a society to maintain, minimize loss of, or maximize gains in welfare). Sensitivity and adaptive capacity, in turn, are composed of sectors (level 3). For adaptive capacity these sectors are human resources, economic capacity, and environmental capacity. For sensitivity, the sectors are settlement/infrastructure, food security, ecosystems, human health, and water resources. Each of these sectors is made up of one to three proxies (level 4). The proxies under adaptive capacity are as follows: human resource proxies are the dependency ratio and literacy rate; economic capacity proxies are GDP (market) per capita and income equity; and environmental capacity proxies are population density, sulfur dioxide divided by state area, and percent of unmanaged land. Proxies in the sensitivity sectors are water availability, fertilizer use per agricultural land area, percent of managed land, life expectancy, birth rate, protein demand, cereal production per agricultural land area, sanitation access, access to safe drinking water, and population at risk from sea level rise.

Each of the hierarchical level values is composed of the geometric means of lower level values. Proxy values are indexed by determining their location within the range of proxy values over all countries or states. The final calculation of resilience is the geometric mean of the adaptive capacity and sensitivity.

Adaptive capacity for a sample of 10 countries from the 160-country study is shown in Figure 7 (base year of 2000). There is a wide range of adaptive capacity represented by these countries. Libya ranks high and Morocco ranks low, both in the sample and overall:

- Russia ranks 32nd and Libya 34th (in the highest quartile).

- Indonesia ranks 45th, Belize 48th, Mexico 59th, and China 75th (in the second quartile).

- The Philippines ranks 91st and India 119th (in the third quartile).

- Morocco ranks 136th and Haiti 156th (in the lowest quartile).

Any country-level analysis must take into account the comparative ranking of the country.

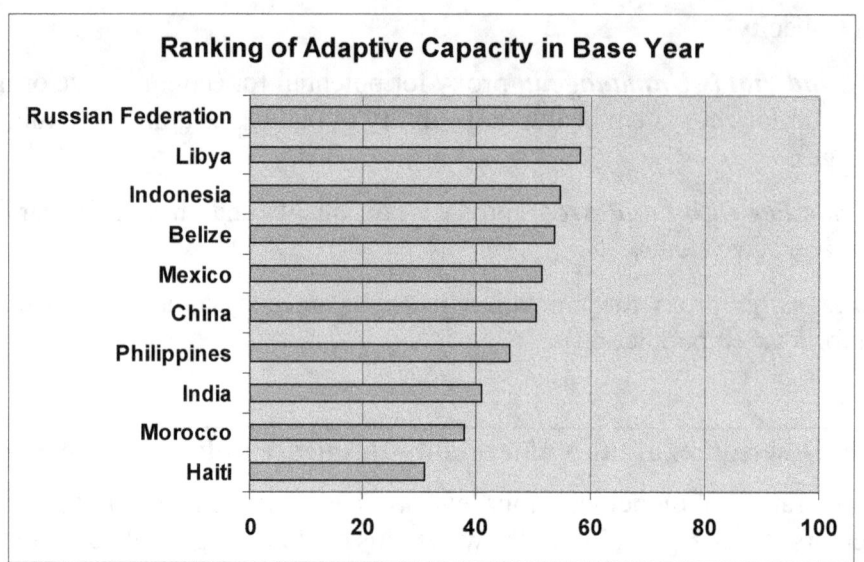

Figure 7. Sample of 10 countries' rankings of adaptive capacity (2000).

Figure 8 shows the contribution of each variable to the overall ranking (slight differences occurring because of the methodology (see box)). In current adaptive capacity, Libya ranks second and Morocco second-to-last among the 10 countries shown in Figure 7. Although Libya has a comparatively unfavorable dependency ratio, it shows comparatively high adaptive capacity in all the environment-related variables. Morocco, on the other hand, ranks poorly on almost all adaptive capacity variables, with the exceptions of emissions and population density.

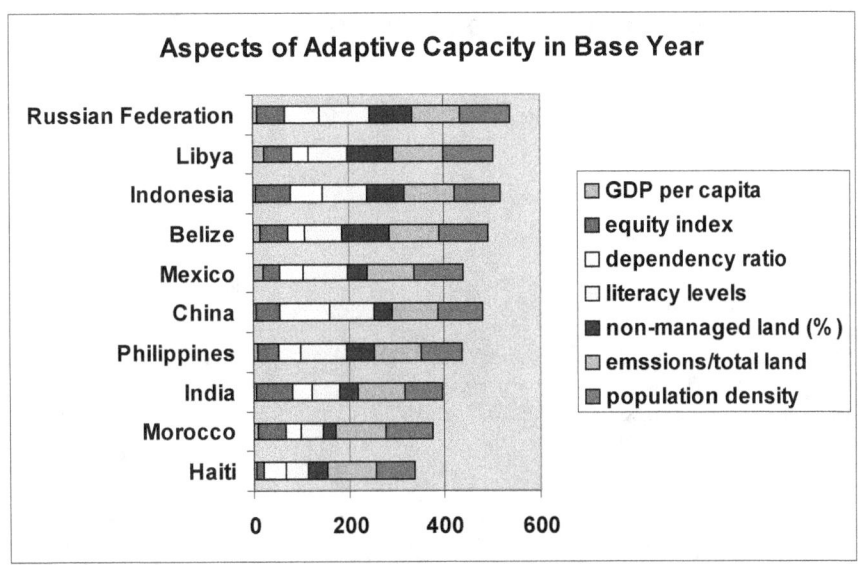

Figure 8. Variables' contributions to adaptive capacity rankings.

Figure 9 shows projected adaptive capacity growth over time for the 10-country sample. Projections are made for two scenarios; rates of growth are based on the IPCC's A1 scenario in its *Special Report on Emissions Scenarios.*[cliv] VRIM simulates two different hypothetical development tracks out to the year 2065 (well beyond the timescale of the present study) with intermediate results at 15-year time steps. These alternative development tracks are not intended to be predictive; they are scenarios.

Both scenarios feature moderate population growth and a tendency toward convergence in affluence (with market-based solutions, rapid technological progress, and improving human welfare). The scenarios used in this study differ in the rate of economic growth, one modeling high-and-fast economic growth, the other delayed growth.

Over time, a low-growth scenario widens the gap among the 10 countries—and the high-growth scenario widens the gap even more. Libya's adaptive capacity over time is almost static in the delayed-growth scenario and increases only slowly in the high-growth scenario. In both scenarios, China and Indonesia outstrip Libya by 2035. Morocco holds its second-to-last place among the ten countries in the sample, but the gap between India and Morocco grows as India is projected to build adaptive capacity at a faster rate.

Figure 10 shows the three categories of adaptive capacity and their contributions to individual countries rankings in North Africa. All countries have relatively low economic capacity and relatively higher environmental capacity. Libya ranks highest in all three aspects among the five countries.

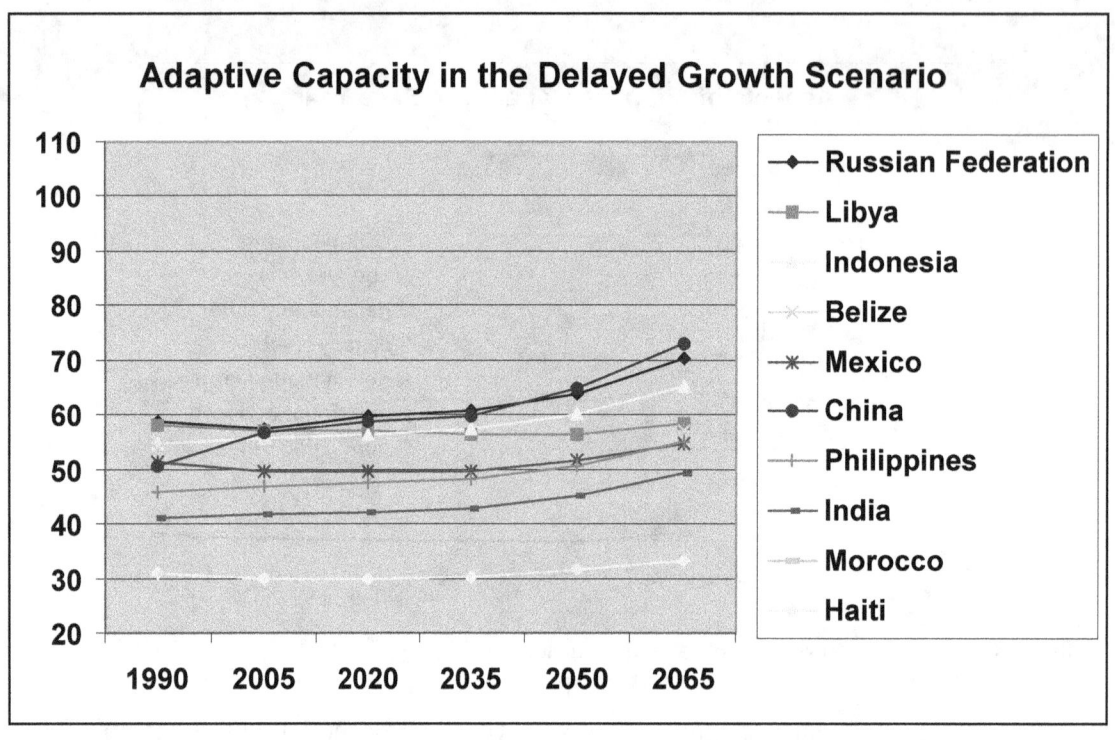

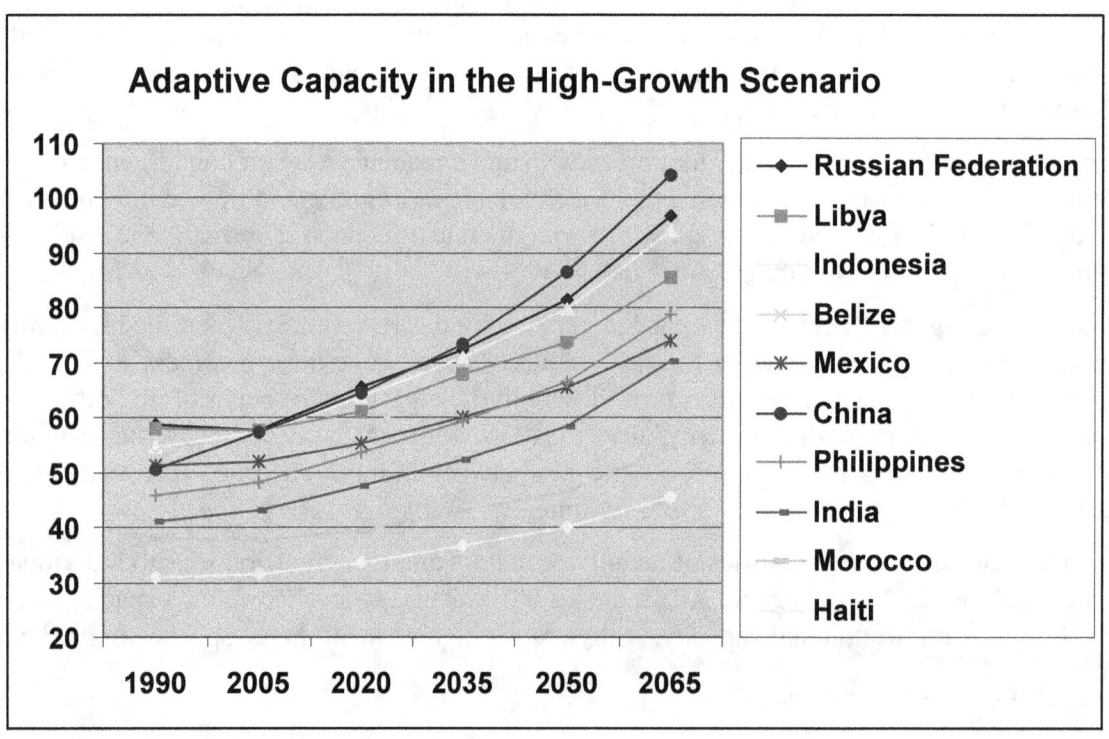

Figure 9. Projections of adaptive capacity for 10 countries.

By 2050 (Figure 11) large differences appear between scenarios, although the relative rankings of the five countries remain the same. Libya makes large gains and the other countries modest gains in adaptive capacity under the high-growth scenario, but all five countries actually lose adaptive capacity by 2050 under the delayed growth scenario.

Thus, assumptions about how conditions unfold in the world and in the region determine rates of (dis)improvement in North Africa's adaptive capacity.

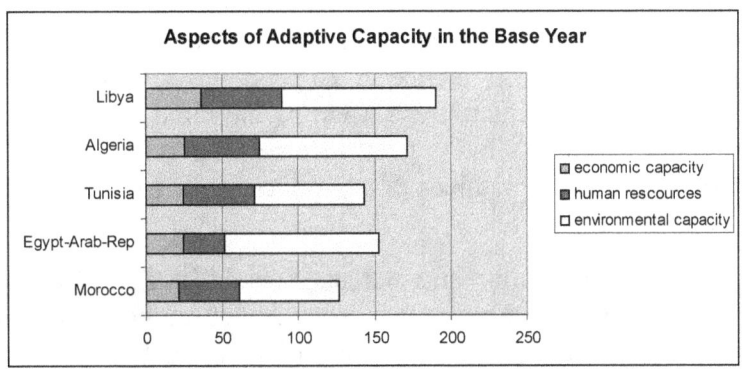

Figure 10. Base year (2000) adaptive capacity of North African countries.

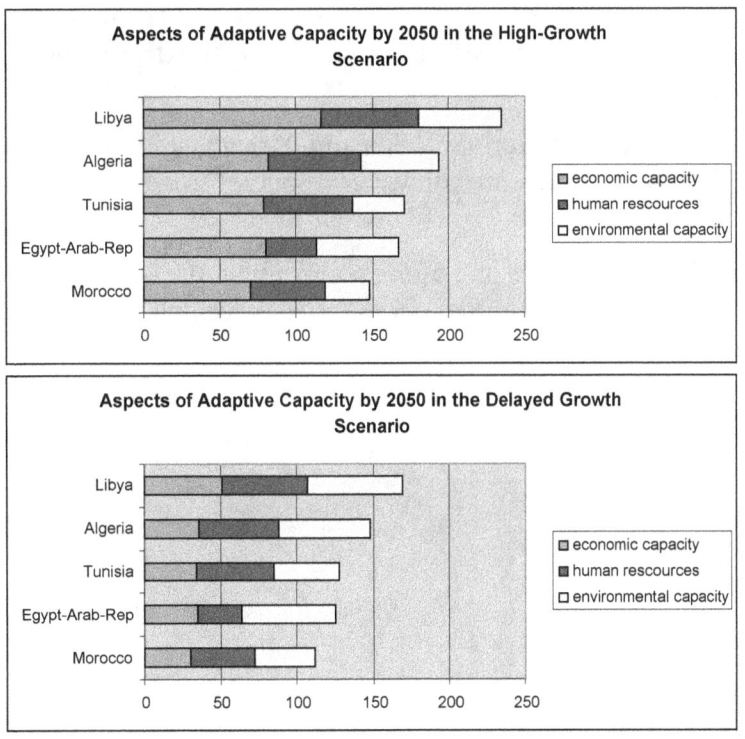

Figure 11. Snapshots of North African adaptive capacity in 2050 under two scenarios

Conclusions: High-Risk Impacts

The sectors in North Africa that are most vulnerable to predicted climate change are the water sector, the agricultural sector, and the coastal zone. Other less studied factors, but clearly existing ones, such as migration, may exacerbate conditions in North Africa in the 21st century.

- *Water Resource Stress.* Water scarcity, even in the absence of climate change, will be one of the most critical problems facing North African countries in the next few decades. It is estimated that Morocco and Algeria's water resources will be reduced by 10-15 percent by 2020, Tunisia's water resources will decline by 28 percent by 2030, and 74.8 percent of Egyptians will have less than adequate fresh water by the same year. Conflicts over water, as have been observed in the past, are likely to surface between African countries. In addition, low-efficiency surface irrigation practices may produce higher water losses, decreases in land productivity, and increased salination.

- *Agriculture.* Reduced annual rainfall and increased persistence and frequency of droughts related to climate change may have negative consequences on the region. Morocco and Tunisia's agricultural sector has already been largely impacted by increasingly frequent droughts. Egypt, where agriculture is impossible without irrigation, is at risk of being largely impacted. However, model results are inconsistent regarding future changes in crop yields and agricultural growing seasons in North Africa. One modeling study suggests that future increases in atmospheric CO_2 concentrations will increase maize yields in Morocco. Nevertheless, as the price of water becomes apparent, North African countries will likely rely more on food imports.

- *Migration.* Increased migration from Sub-Saharan Africa and increased internal migration to urban areas will place further pressures on water resources, food availability, infrastructure and the ecosystem.

- *Natural Disasters.* North Africa is one of the regions with the largest predicted impacts from flooding given its close proximity to the Mediterranean Sea and the highly concentrated and poor populations along the coasts.

Annex A:

Accuracy of Regional Models

Below is an excerpt from IPCC (2007), Chapter 11, Regional models; see IPCC 2007 for references. [8]

11.2.2 Skill of Models in Simulating Present and Past Climates

There are biases in the simulations of African climate that are systematic across the multi-model dataset (MMD) models, with 90% of models overestimating precipitation in southern Africa, by more than 20% on average (and in some cases by as much as 80%) over a wide area often extending into equatorial Africa. The temperature biases over land are not considered large enough to directly affect the credibility of the model projections.

The Inter-Tropical Convergence Zone (ITCZ) in the Atlantic is displaced equatorward in nearly all of these AOGCM [Atmosphere-Ocean General Circulation Model] simulations. Ocean temperatures are too warm by an average of 1°C to 2°C in the Gulf of Guinea and typically by 3°C off the southwest coast in the region of intense upwelling, which is clearly too weak in many models. In several of the models there is no West African monsoon as the summer rains fail to move from the Gulf onto land, but most of the models do have a monsoonal climate albeit with some distortion. Moderately realistic interannual variability of sea surface temperatures (SSTs) in the Gulf of Guinea and the associated dipolar rainfall variations in the Sahel and the Guinean Coast are, by the criteria of Cook and Vizy (2006), only present in 4 of the 18 models examined. Tennant (2003) describes biases in several AGCMs [Atmospheric General Circulation Models], such as the equatorward displacement of the mid-latitude jet in austral summer, a deficiency that persists in the most recent simulations.

Despite these deficiencies, AGCMs can simulate the basic pattern of rainfall trends in the second half of the 20th century if given the observed SST evolution as boundary conditions, as described in the multi-model analysis of Hoerling et al (2006) and the growing literature on the interannual variability and trends in individual models (e.g., Rowell et al, 1995; Bader and Latif, 2003; Giannini et al, 2003; Haarsma et al, 2005; Kamga et al, 2005; Lu and Delworth, 2005). However, there is less confidence in the ability of AOGCMs to generate interannual variability in the SSTs of the type known to affect African rainfall, as evidenced by the fact that very few AOGCMs produce droughts comparable in magnitude to the Sahel drought of the 1970s and 1980s (Hoerling et al, 2006). There are exceptions, but what distinguishes these from the bulk of the models is not understood.

The very wet Sahara 6 to 8 ka is thought to have been a response to the increased summer insolation due to changes in the Earth's orbital configuration. Modelling studies of this response provide background information on the quality of a model's African monsoon, but the processes controlling the response to changing seasonal insolation may be different from those controlling the response to increasing greenhouse gases. The fact that GCMs have difficulty in simulating the full magnitude of the mid-Holocene wet period, especially in the absence of vegetation feedbacks, may indicate a lack of sensitivity to other kinds of forcing (Jolly et al, 1996; Kutzbach et al, 1996).

---

[8] Some references in this section have been changed to be internally consistent with this document and other references have been removed to avoid confusion.

Regional climate modelling has mostly focused on southern Africa, where the models generally improve on the climate simulated by global models but also share some of the biases in the global models. For example, Engelbrecht et al (2002) and Arnell et al (2003) both simulate excessive rainfall in parts of southern Africa, reminiscent of the bias in the MMD. Hewitson et al (2004) and Tadross et al (2006) note strong sensitivity to the choice of convective parametrization, and to changes in soil moisture and vegetative cover (New et al, 2003; Tadross et al, 2005a), reinforcing the view (Rowell et al, 1995) that land surface feedbacks enhance regional climate sensitivity over Africa's semi-arid regions. Over West Africa, the number of Regional Climate Model (RCM) investigations is even more limited (Jenkins et al, 2002; Vizy and Cook, 2002). The quality of the 25-year simulation undertaken by Paeth et al (2005) is encouraging, emphasizing the role of regional SSTs and changes in the land surface in forcing West African rainfall anomalies. Several recent AGCM time-slice simulations focusing on tropical Africa show good simulation of the rainy season (Coppola and Giorgi, 2005; Caminade et al, 2006; Oouchi et al, 2006).

Hewitson and Crane (2005) developed empirical downscaling for point-scale precipitation at sites spanning the continent, as well as a 0.1° resolution grid over South Africa. The downscaled precipitation forced by reanalysis data provides a close match to the historical climate record, including regions such as the eastern escarpment of the sub-continent that have proven difficult for RCMs.

Annex B:

Knowledge Deficiencies that Preclude a Full Evaluation of Climate Change Impacts on North Africa and North Africa's Adaptive Strategies

To increase the likelihood that this evaluation represents a reasonable assessment of North Africa's projected climate changes and their impacts, as well as the region's adaptive capacity, the following gaps would need to be addressed:

- In physical science research, regional analyses will continue to be limited by the inability to model regional climates satisfactorily, including complexities arising from the interaction of global, regional, and local processes. Uncertainties in changing precipitation, dust storms, and desertification leave important gaps in knowledge needed for climate projections. One gap of particular interest is the lack of medium-term (20 to 30 year) projections that could be relied upon for planning purposes. Similarly, scientific projections of water supply and agricultural productivity are limited by inadequate understanding of various climate and physical factors affecting both areas. Similar types of issues exist for the biological and ecological systems that are affected. These gaps are particularly acute for North Africa, where regional models are poor and relatively few impacts studies have been conducted.

- In social science research, scientists and analysts have only a partial understanding of the important factors in vulnerability, resilience, and adaptive capacity—much less their interactions and evolution. Again, research agendas on vulnerability, adaptation, and decisionmaking abound (e.g., (http://books.nap. edu/catalog.php?record_id=12545).

- Important factors are unaccounted for in research; scientists know what some of them are, but there are likely factors whose influence will be surprising. An example from earlier research on the carbon cycle illustrates this situation. The first carbon cycle models did not include carbon exchanges involving the terrestrial domain. Modelers assumed that the exchange was about equal and the only factor modeled was deforestation. This assumption, of course, made the models inadequate for their purposes. In another example, ecosystems research models are only beginning to account for changes in pests, e.g., the pine bark beetle.

- Social models or parts of models in climate research have been developed to simulate consumption (with the assumption of well-functioning markets and rational actor behavior) and mitigation/adaptation policies (but without attention to the social feasibility of enacting or implementing such policies). As anthropogenic climate change is the result of human decisions, the lack of knowledge about motivation, intent, and behavior is a serious shortcoming.

Overall, research about climate change impacts on North Africa has been undertaken piecemeal: discipline by discipline, sector by sector, with political implications separately considered from physical effects. This knowledge gap can be remedied by integrated research into energy-economic-environmental-political conditions and possibilities.

i According to the *Encyclopedia of Nations*, the high unemployment rate, also estimated for 2000, is linked with the country's years of sanctions, which have affected Libya's oil and gas exports, in addition to Mu'ammar al-Qadhafi's efforts at preventing the emergence of the growing private sector. Source: *Encyclopedia of the Nations*, s.v. "Libya Working Conditions," http://www.nationsencyclopedia.com/economies/Africa/Libya-WORKING-CONDITIONS.html (accessed May 28, 2009).

ii Central Intelligence Agency (CIA), *The 2008 World Factbook*, https://www.cia.gov/library/publications/the-world-factbook/index html (accessed May 15, 2009).

iii Central Intelligence Agency (CIA), *The 2008 World Factbook*, https://www.cia.gov/library/publications/the-world-factbook/index html (accessed May 15, 2009).

iv Weather Online, *Wind of the World: Sorocco*, http://www.weatheronline.co.uk/reports/wind/The-Sirocco htm (accessed May 15, 2009).

v Intergovernmental Panel on Climate Change (IPCC), *Climate Change 2007: the Physical Science Basis*, eds. S. Solomon, D. Qin, M. Manning, M. Marquis, K. Averyt, M.M.B. Tignor, H.L. Jr. Miller, and Z. Chen (Cambridge: Cambridge University Press, 2007), http://www.ipcc.ch/ipccreports/ar4-wg1 htm.

vi Intergovernmental Panel on Climate Change (IPCC), *Special Report on Emissions Scenarios (SRES)*, eds. Nebojsa Nakicenovic and Rob Swart (Cambridge: Cambridge University Press, 2000), http://www.ipcc.ch/ipccreports/sres/emission/index htm.

vii Intergovernmental Panel on Climate Change (IPCC), *The IPCC Data Distribution Centre, HadCM2 GCM Model Information*, http://www.ipcc-data.org/is92/hadcm2_info.html (accessed April 1, 2009).

viii Intergovernmental Panel on Climate Change (IPCC), *The IPCC Data Distribution Centre, HadCM2 GCM Model Information*, http://www.ipcc-data.org/is92/hadcm2_info.html (accessed April 1, 2009).

ix M. Hulme, R. Doherty, T. Ngara, M. New, and D. Lister, "African Climate Change: 1900-2100." *Climate Research* 17, no. 2 (2001): 145-68.

x M. Hulme, R. Doherty, T. Ngara, M. New, and D. Lister, "African Climate Change: 1900-2100." *Climate Research* 17, no. 2 (2001): 145-68; H. Paeth and H. P. Thamm, "Regional Modelling of Future African Climate North of 15 °S Including Greenhouse Warming and Land Degradation." *Climatic Change* 83, no. 3 (2007): 401-27; and H. Paeth, K. Born, R. Girmes, R. Podzun, and D. Jacob, "Regional Climate Change in Tropical and Northern Africa Due to Greenhouse Forcing and Land Use Changes." *Journal of Climate* 22, no. 1 (2009): 114-32.

xi H. Paeth, K. Born, R. Girmes, R. Podzun, and D. Jacob, "Regional Climate Change in Tropical and Northern Africa Due to Greenhouse Forcing and Land Use Changes." *Journal of Climate* 22, no. 1 (2009): 114-32.

xii Intergovernmental Panel on Climate Change (IPCC), *Climate Change 2007: the Physical Science Basis*, eds. S. Solomon, D. Qin, M. Manning, M. Marquis, K. Averyt, M.M.B. Tignor, H.L. Jr. Miller, and Z. Chen (Cambridge: Cambridge University Press, 2007), http://www.ipcc.ch/ipccreports/ar4-wg1 htm.

xiii Intergovernmental Panel on Climate Change (IPCC), *Climate Change 2007: the Physical Science Basis*, eds. S. Solomon, D. Qin, M. Manning, M. Marquis, K. Averyt, M.M.B. Tignor, H.L. Jr. Miller, and Z. Chen (Cambridge: Cambridge University Press, 2007), http://www.ipcc.ch/ipccreports/ar4-wg1 htm.

xiv Intergovernmental Panel on Climate Change (IPCC), *Climate Change 2007: the Physical Science Basis*, eds. S. Solomon, D. Qin, M. Manning, M. Marquis, K. Averyt, M.M.B. Tignor, H.L. Jr. Miller, and Z. Chen (Cambridge: Cambridge University Press, 2007), http://www.ipcc.ch/ipccreports/ar4-wg1 htm.

xv Intergovernmental Panel on Climate Change (IPCC), *Climate Change 2007: the Physical Science Basis*, eds. S. Solomon, D. Qin, M. Manning, M. Marquis, K. Averyt, M.M.B. Tignor, H.L. Jr. Miller, and Z. Chen (Cambridge: Cambridge University Press, 2007), http://www.ipcc.ch/ipccreports/ar4-wg1 htm.

xvi Intergovernmental Panel on Climate Change (IPCC), *Climate Change 2007: the Physical Science Basis*, eds. S. Solomon, D. Qin, M. Manning, M. Marquis, K. Averyt, M.M.B. Tignor, H.L. Jr. Miller, and Z. Chen (Cambridge: Cambridge University Press, 2007), http://www.ipcc.ch/ipccreports/ar4-wg1 htm.

xvii M. Hulme, R. Doherty, T. Ngara, M. New, and D. Lister, "African Climate Change: 1900-2100." *Climate Research* 17, no. 2 (2001): 145-68.

xviii A. L. Gibelin and M. Déqué, "Anthropogenic Climate Change over the Mediterranean Region Simulated by a Global Variable Resolution Model." *Climate Dynamics* 20, no. 4 (2003): 327-39.

xix F. Giorgi, "Climate Change Hot-Spots." *Geophysical Research Letters* 33, no. 8 (2006).

xx H. Paeth, K. Born, R. Girmes, R. Podzun, and D. Jacob, "Regional Climate Change in Tropical and Northern Africa Due to Greenhouse Forcing and Land Use Changes." *Journal of Climate* 22, no. 1 (2009): 114-32.

[xxi] A. Chaponniere and V. Smakhtin," A Review of Climate Change Scenarios and Preliminary Rainfall Trend Analysis in the Oum Er Rbia Basin, Morocco," Colombo, Sri Lanka: (IWMI working paper 110: Drought series, paper 8) p. 16; and Ministère de l'Aménagement du Territoire, de l'Urbanisme, de l'Habitat et de l'Environnement, "Communication Nationale Initiale À La Convention Cadre Des Nations Unies Sur Les Changements Climatiques (First National Communication to the United Nations Framework Convention on Climate Change)," Kingdom of Morocco, October, 2001, http://unfccc.int/resource/docs/natc/mornc1e.pdf.

[xxii] M.N. Tsimplis and N.E. Spencer, "Collection and analysis of monthly mean sea level data in the Mediterranean and the Black Sea." *Journal of Coastal Research* 13, no.2 (1997): 534–544.

[xxiii] J.A. Church, N.J. White, R. Coleman, K. Lambeck, and J.X. Mitrovica, "Estimates of regional distribution of sea level rise over the 1950-2000 period." *Journal of Climate* 17 (2004): 2609–2625.

[xxiv] Intergovernmental Panel on Climate Change (IPCC), *Climate Change 2007: the Physical Science Basis*, eds. S. Solomon, D. Qin, M. Manning, M. Marquis, K. Averyt, M.M.B. Tignor, H.L. Jr. Miller, and Z. Chen (Cambridge: Cambridge University Press, 2007), http://www.ipcc.ch/ipccreports/ar4-wg1 htm.

[xxv] Intergovernmental Panel on Climate Change (IPCC), *Climate Change 2007: the Physical Science Basis*, eds. S. Solomon, D. Qin, M. Manning, M. Marquis, K. Averyt, M.M.B. Tignor, H.L. Jr. Miller, and Z. Chen (Cambridge: Cambridge University Press, 2007), http://www.ipcc.ch/ipccreports/ar4-wg1 htm.

[xxvi] S. Somot, F. Sevault, M. Deque, and M. Crepon, "21st Century Climate Change Scenario for the Mediterranean Using a Coupled Atmosphere-Ocean Regional Climate Model." *Global and Planetary Change* 63, no. 2-3 (2008): 112-26.

[xxvii] Intergovernmental Panel on Climate Change (IPCC), *Climate Change 2007: the Physical Science Basis*, eds. S. Solomon, D. Qin, M. Manning, M. Marquis, K. Averyt, M.M.B. Tignor, H.L. Jr. Miller, and Z. Chen (Cambridge: Cambridge University Press, 2007), http://www.ipcc.ch/ipccreports/ar4-wg1 htm.

[xxviii] Michael N. Tsimplis, Marta Marcos, Samuel Somot, "21st century Mediterranean sea level rise: Steric and atmospheric pressure contributions from a regional model." *Global and Planetary Change*, 63 (2008), 105-111.

[xxix] Intergovernmental Panel on Climate Change (IPCC), *Climate Change 2007: the Physical Science Basis*, eds. S. Solomon, D. Qin, M. Manning, M. Marquis, K. Averyt, M.M.B. Tignor, H.L. Jr. Miller, and Z. Chen (Cambridge: Cambridge University Press, 2007), http://www.ipcc.ch/ipccreports/ar4-wg1 htm.

[xxx] M. Snoussi, T. Ouchani, and S. Niazi, "Vulnerability Assessment of the Impact of Sea-Level Rise and Flooding on the Moroccan Coast: The Case of the Mediterranean Eastern Zone." *Estuarine Coastal and Shelf Science* 77, no. 2 (2008): 206-13.

[xxxi] R.A Warrick, C. Le Provost, M.F. Meier, J. Oerlemans, P.L. Woodworth, "Changes in sea level," *The Science of Climate Change*, eds. Houghton, J.T., Meira Filho, L.G., Callander, B.A., Harris, N., Kattenberg, A., Maskell, K. (Cambridge: Cambridge University Press, 1996).

[xxxii] Intergovernmental Panel on Climate Change (IPCC), *Climate Change 2007: the Physical Science Basis*, eds. S. Solomon, D. Qin, M. Manning, M. Marquis, K. Averyt, M.M.B. Tignor, H.L. Jr. Miller, and Z. Chen (Cambridge: Cambridge University Press, 2007), http://www.ipcc.ch/ipccreports/ar4-wg1 htm.

[xxxiii] Ministère De L'Environnement et de L'Amenagement du Territoir (Ministry of Environment and Land Planning), "Communication Initiale De La Tunisie À La Convention Cadre Des Nations Unies Sur Les Changements Climatiques (Initial Communication of Tunisia to the United Nations Framework Convention on Climate Change)," Republic of Tunisia, 2001, http://unfccc.int/resource/docs/natc/tunnc1esum.pdf.

[xxxiv] A. Agoumi, "Vulnerability of North African Countries to Climatic Changes: Adaptation and Implementation Strategies for Climate Change," Climate Change Knowledge Network, 2003, http://www.cckn net/compendium/north_africa.asp.

[xxxv] A. Agoumi, "Vulnerability of North African Countries to Climatic Changes: Adaptation and Implementation Strategies for Climate Change," Climate Change Knowledge Network, 2003, http://www.cckn net/compendium/north_africa.asp.

[xxxvi] Intergovernmental Panel on Climate Change (IPCC), *Climate Change 2007: the Physical Science Basis*, eds. S. Solomon, D. Qin, M. Manning, M. Marquis, K. Averyt, M.M.B. Tignor, H.L. Jr. Miller, and Z. Chen (Cambridge: Cambridge University Press, 2007), http://www.ipcc.ch/ipccreports/ar4-wg1 htm.

[xxxvii] Ministère de l'Aménagement du Territoire, de l'Urbanisme, de l'Habitat et de l'Environnement, "Communication Nationale Initiale À La Convention Cadre Des Nations Unies Sur Les Changements Climatiques (First National Communication to the United Nations Framework Convention on Climate Change)," Kingdom of Morocco, October, 2001, http://unfccc.int/resource/docs/natc/mornc1e.pdf.

[xxxviii] R. J. Nicholls, "Coastal Flooding and Wetland Loss in the 21st Century: Changes under the SRES Climate and Socio-Economic Scenarios." *Global Environmental Change-Human and Policy Dimensions* 14, no. 1 (2004): 69-86.

[xxxix] M. Snoussi, T. Ouchani, and S. Niazi, "Vulnerability Assessment of the Impact of Sea-Level Rise and Flooding on the Moroccan Coast: The Case of the Mediterranean Eastern Zone." *Estuarine Coastal and Shelf Science* 77, no. 2 (2008): 206-13.

[xl] Abigail Somma, "Squeezing the Most out of Scarce Water Resources,"*IFPRI Forum*, I (2009), 10.

[xli] FAOSTAT, s.v., "Land use 2005," http://faostat fao.org/.

[xlii] Josephine Khaoma W. Ngaira, "Impact of climate change on agriculture in Africa by 2030." *Scientific Research and Essays* 2, no. 7 (July 2007): 238-243; and M. Kassas, "Agriculture in North Africa: Sociocultural Aspects," *Journal of Agricultural Ethics* 2 (1989): 183-190.

[xliii] Encyclopedia of the Nation, http://www.nationsencyclopedia.com (accessed June 9, 2009).

[xliv] P.J. Ashton, "Avoiding conflicts over Africa's water resources," *Ambio* 31 (2002): 236.

[xlv] L. Gueye et al (2005), reported in M. Boko, I. Niang, A. Nyong and C. Vogel, "Africa," in IPCC [Intergovernmental Panel on Climate Change]. *Climate Change 2007: Impacts, Adaptation and Vulnerability*, eds. Martin Parry, Osvaldo Canziani, Jean Palutikof, Paul van der Linden, and Clair Hanson (Cambridge: Cambridge University Press, 2007).

[xlvi] IPCC [Intergovernmental Panel on Climate Change]. *Climate Change 2007: Impacts, Adaptation and Vulnerability*, eds. Martin Parry, Osvaldo Canziani, Jean Palutikof, Paul van der Linden, and Clair Hanson (Cambridge: Cambridge University Press, 2007)

[xlvii] FAO (Food and Agriculture Organization of the United Nations), "Ministerial Conference on Water for Agriculture and Energy in Africa: The Challenges of Climate Change" (conference, Irrigation Projections for 2030-2050, Sirte, Libyan Arab Jamahiriya, December 15-17, 2008).

[xlviii] R. Schubert, H.J. Schellnhuber, N. Buchmann, A. Epiney, R. GrieBhammer, M. Kulessa, D. Messner, S. Rahmstorf, J. Schmid. *Climate Change as a Security Risk*, German Advisory Council on Global Change (WBGU), trans. Christopher Hay and Seeheim-Jugenheim (London and Sterling, VA: Earthscan, 2008).

[xlix] FAONewsroom of the United Nations Food and Agriculture Organization, "Agriculture in the Near East likely to suffer from climate change: The hungry and poor will be most affected – FAO meeting debates impact on the region," (March 3, 2008), http://www fao.org/newsroom/en/news/2008/1000800/index.html.

[l] *Climate Change and its Impact on Health in Morocco*, (Amman, Jordan: Presentation to the "Regional Workshop on Adaptation Strategies to Protect Health under Climate Change Variability and Change in Water Stressed Countries in EMR," December 11-14, 2006), slides, www.emro.who.int/Ceha/media/powerpoint/morocco%20CP.pps (accessed March 26, 2009)

[li] IPCC [Intergovernmental Panel on Climate Change]. *Climate Change 2007: Impacts, Adaptation and Vulnerability*, eds. Martin Parry, Osvaldo Canziani, Jean Palutikof, Paul van der Linden, and Clair Hanson (Cambridge: Cambridge University Press, 2007)

[lii] IPCC [Intergovernmental Panel on Climate Change]. *Climate Change 2007: Impacts, Adaptation and Vulnerability*, eds. Martin Parry, Osvaldo Canziani, Jean Palutikof, Paul van der Linden, and Clair Hanson (Cambridge: Cambridge University Press, 2007).

[liii] Climate Protection Programme (GTZ), "Development of a Strategy for Adaptation to Climate Change in the Tunisian Agricultural Sector," Factsheet, 2007, http://www.gtz.de/de/dokumente/gtz-en-anpassung-klimawandel-tunesien.pdf (accessed March 26, 2009).

[liv] D. Moran "Climate Change and Regional Security" (workshop, Department of National Security Affairs at the Naval Postgraduate School, sponsored by Long Range Assessment Unit of the National Intelligence Council, Monterey, California, December 11-13, 2007).

[lv] Irna van der Molen and Antoinette Hildering, "Water: Cause for conflict or co-operation?" *Journal on Science and World Affairs* 1, no. 2 (2005): 133-143.

[lvi] Ragab Ragab and Christel Prudhomme, "Climate change and water resources management in arid and semi-arid regions: prospective and challenges for the 21st century." *Biosyst. Eng* 81 (2002): 3–34, doi:10.1006/bioe.2001.0013.

[lvii] Ragab Ragab and Christel Prudhomme, "Climate change and water resources management in arid and semi-arid regions: prospective and challenges for the 21st century." *Biosyst. Eng* 81 (2002): 3–34, doi:10.1006/bioe.2001.0013.

[lviii] Egyptian Environmental Affairs Agency, "The Arab Republic of Egypt: Initial National Communication on Climate Change," Prepared for the United Nations Framework Convention on Climate Change (UNFCCC), June 1999), http://unfccc.int/resource/docs/natc/egync1.pdf.

[lix] Alan Richards and John Waterbury. *A Political Economy of the Middle East*, 3rd ed. (Boulder, CO: Westview Press, 2007), 174-175.

[lx] Alan Richards and John Waterbury. *A Political Economy of the Middle East*, 3rd ed. (Boulder, CO: Westview Press, 2007), 174-175; and Ronald Bleier, "Will Nile Water Go to Israel?: North Sinai Pipelines and the Politics of Scarcity." *Middle East Policy*, V, no. 3 (September 1997): 113-124.

[lxi] Ragab Ragab and Christel Prudhomme, "Climate change and water resources management in arid and semi-arid regions: prospective and challenges for the 21st century." *Biosyst. Eng* 81 (2002): 3–34, doi:10.1006/bioe.2001.0013.

[lxii] Ragab Ragab and Christel Prudhomme, "Climate change and water resources management in arid and semi-arid regions: prospective and challenges for the 21st century." *Biosyst. Eng* 81 (2002): 3–34, doi:10.1006/bioe.2001.0013.

[lxiii] Ragab Ragab and Christel Prudhomme, "Climate change and water resources management in arid and semi-arid regions: prospective and challenges for the 21st century." *Biosyst. Eng* 81 (2002): 3–34, doi:10.1006/bioe.2001.0013.

[lxiv] http://unfccc.int/resource/docs/natc/egync1.pdf

[lxv] IPCC [Intergovernmental Panel on Climate Change]. *Climate Change 2007: Impacts, Adaptation and Vulnerability*, eds. Martin Parry, Osvaldo Canziani, Jean Palutikof, Paul van der Linden, and Clair Hanson (Cambridge: Cambridge University Press, 2007).

[lxvi] Julia Bucknall. *Making the most of Scarcity: Accountability for the Better Water Management in the Middle East and North Africa* (Washington D.C.: World Bank, 2007); and World Bank, "Coping with Scarce Water in the Middle East and North Africa" http://go.worldbank.org/4F8D9IG4N0 (accessed June 2, 2009).

[lxvii] Julia Bucknall. *Making the most of Scarcity: Accountability for the Better Water Management in the Middle East and North Africa* (Washington D.C.: World Bank, 2007).

[lxviii] Dennis Wichelns, "The role of 'virtual water' in efforts to achieve food security and other national goals, with an example from Egypt," *Agricultural Water Management* 49 (2001): 131-151.

[lxix] Ayman F. Abou-Hadid, "Assessment of Impacts, Adaptation and Vulnerability to Climate Change in North Africa: Food Production and Water Resources" Central Laboratory for Agricultural Climate (CLAC) at the Agricultural Research Centre of the Ministry of Agriculture and Land Reclamation, Egypt (The International START Secretariat: Washington, DC, 2006).

[lxx] FAO (Food and Agriculture Organization of the United Nations), "Ministerial Conference on Water for Agriculture and Energy in Africa: The Challenges of Climate Change" (conference, Irrigation Projections for 2030-2050, Sirte, Libyan Arab Jamahiriya, December 15-17, 2008).

[lxxi] Ragab Ragab and Christel Prudhomme, "Climate change and water resources management in arid and semi-arid regions: prospective and challenges for the 21st century." *Biosyst. Eng* 81 (2002): 3–34, doi:10.1006/bioe.2001.0013.

[lxxii] Lasse Ringius, Thomas E. Downing, Mike Hulme, Dominic Waughray, Rolf Selrod, "Climate Change in Africa - Issues and Challenges in Agriculture and Water for Sustainable Development," ISSN: 0804-4562, Center for International Climate and Environmental Research – Oslo (CICERO) (Oslo: University of Oslo, November 1996).

[lxxiii] D. N. Yates and K. M. Strzepek, "An Assessment of Integrated Climate Change Impacts on the Agricultural Economy of Egypt." *Climatic Change* 38, no. 3 (1998): 261-87.

[lxxiv] German Advisory Council on Global Change (WBGU), *Climate Change as a Security Risk*, eds. R. Schubert, H.J. Schellnhuber, N. Buchmann, A. Epiney, R. Grießhammer, M. Kulessa, D. Messner, S. Rahmstorf, and J. Schmid (London: Earthscan , 2008).

[lxxv] FAONewsroom, Agriculture in the Near East likely to suffer from climate change: The hungry and poor will be most affected – FAO meeting debates impact on the region, March 3, 2008, Rome/Cairo, http://www.fao.org/newsroom/en/news/2008/1000800/index.html.

[lxxvi] D. Moran "Climate Change and Regional Security" (workshop, Department of National Security Affairs at the Naval Postgraduate School, sponsored by Long Range Assessment Unit of the National Intelligence Council, Monterey, California, December 11-13, 2007).

[lxxvii] Ministère de l'Aménagement du Territoire, de l'Urbanisme, de l'Habitat et de l'Environnement, "Communication Nationale Initiale À La Convention Cadre Des Nations Unies Sur Les Changements Climatiques (First National Communication to the United Nations Framework Convention on Climate Change)," Kingdom of Morocco, October, 2001, http://unfccc.int/resource/docs/natc/mornc1e.pdf.

[lxxviii] P. G. Jones and P. K. Thornton, "The Potential Impacts of Climate Change on Maize Production in Africa and Latin America in 2055." *Global Environmental Change-Human and Policy Dimensions* 13, no. 1 (2003): 51-59.

[lxxix] Steven J. Crafts-Brandner and Michael E. Salvucci, "Sensitivity of Photosynthesis in a C4 Plant, Maize, to Heat Stress." *Plant Physiology* 129 (2002): 1773–1780.

49

[lxxx] Helmy M. Eid, Samia M. El-Marsafawy, and Samiha A. Ouda, "Assessing the Economic Impacts of Climate Change on Agriculture in Egypt: A Ricardian Approach" Policy Research Working Paper (Development Research Group of The World Bank, July 2007), http://www-wds.worldbank.org/external/default/WDSContentServer/IW3P/IB/2007/07/31/000158349_20070731143402/Rendered/PDF/wps4293.pdf (accessed April 6, 2009); and Ragab Ragab and Christel Prudhomme, "Climate change and water resources management in arid and semi-arid regions: prospective and challenges for the 21st century." *Biosyst. Eng* 81 (2002): 3–34, doi:10.1006/bioe.2001.0013.

[lxxxi] Josephine Khaoma W. Ngaira, "Impact of climate change on agriculture in Africa by 2030." *Scientific Research and Essays* 2, no. 7 (July 2007): 238-243.

[lxxxii] Egypt State Information Service: Your Gateway to Egypt, s.v. "Agriculture," http://www.sis.gov.eg/En/Economy/Sectors/Agriculture/050301000000000001.htm (accessed April 6, 2009).

[lxxxiii] Egypt State Information Service: Your Gateway to Egypt, s.v. "Agriculture," http://www.sis.gov.eg/En/Economy/Sectors/Agriculture/050301000000000001.htm (accessed April 6, 2009).

[lxxxiv] Ayman F. Abou-Hadid, "Assessment of Impacts, Adaptation, and Vulnerability to Climate Change in North Africa: Food Production and Water Resources" A Final Report Submitted to Assessments of Impacts and Adaptations to Climate Change (AIACC), Project No. AF 90 (Washington, D.C.: International START Secretariat, 2006).

[lxxxv] Ayman F. Abou-Hadid, "Assessment of Impacts, Adaptation, and Vulnerability to Climate Change in North Africa: Food Production and Water Resources" A Final Report Submitted to Assessments of Impacts and Adaptations to Climate Change (AIACC), Project No. AF 90 (Washington, D.C.: International START Secretariat, 2006).

[lxxxvi] H.M. El-Shaer, C. Rosenzweig, A. Iglesias, M.H. Eid, and D. Hillel, "Impact of Climate Change on Possible Scenarios for Egyptian Agriculture in the Future." *Mitigation and Adaptation Strategies for Global Change* 1 (1997): 233-50.

[lxxxvii] P. G. Jones and P. K. Thornton, "The Potential Impacts of Climate Change on Maize Production in Africa and Latin America in 2055." *Global Environmental Change-Human and Policy Dimensions* 13, no. 1 (2003): 51-59.

[lxxxviii] H.M. El-Shaer, C. Rosenzweig, A. Iglesias, M.H. Eid, and D. Hillel, "Impact of Climate Change on Possible Scenarios for Egyptian Agriculture in the Future." *Mitigation and Adaptation Strategies for Global Change* 1 (1997): 233-50.

[lxxxix] D. N. Yates and K. M. Strzepek, "An Assessment of Integrated Climate Change Impacts on the Agricultural Economy of Egypt." *Climatic Change* 38, no. 3 (1998): 261-87.

[xc] Intergovernmental Panel on Climate Change (IPCC), *Climate Change 2007: the Physical Science Basis*, eds. S. Solomon, D. Qin, M. Manning, M. Marquis, K. Averyt, M.M.B. Tignor, H.L. Jr. Miller, and Z. Chen (Cambridge: Cambridge University Press, 2007), http://www.ipcc.ch/ipccreports/ar4-wg1 htm.

[xci] H.M. El-Shaer, C. Rosenzweig, A. Iglesias, M.H. Eid, and D. Hillel, "Impact of Climate Change on Possible Scenarios for Egyptian Agriculture in the Future." *Mitigation and Adaptation Strategies for Global Change* 1 (1997): 233-50.

[xcii] A. K. Hegazy, M. A. Medany, H. F. Kabiel, and M. M. Maez, "Spatial and Temporal Projected Distribution of Four Crop Plants in Egypt." *Natural Resources Forum* 32, no. 4 (2008): 316-26.

[xciii] Helmy M. Eid, Samia M. El-Marsafawy, and Samiha A. Ouda, "Assessing the Economic Impacts of Climate Change on Agriculture in Egypt: A Ricardian Approach" Policy Research Working Paper (Development Research Group of The World Bank, July 2007), http://www-wds.worldbank.org/external/default/WDSContentServer/IW3P/IB/2007/07/31/000158349_20070731143402/Rendered/PDF/wps4293.pdf (accessed April 6, 2009).

[xciv] The Ricardian approach is based on the following hypothesis: (1) Climate shifts the production function for crops; (2) There is perfect competition in both product and input prices; (3) The land values have attained the long-run equilibrium associated with each region climate; (4) Market prices are unchanged as a result of change in environmental conditions; (5) Adaptation takes place by all means including the adoption of new crops or farming systems; and (6) The adaptation cost is not considered in the analysis. Source: Helmy M. Eid, Samia M. El-Marsafawy, and Samiha A. Ouda, "Assessing the Economic Impacts of Climate Change on Agriculture in Egypt: A Ricardian Approach" Policy Research Working Paper (Development Research Group of The World Bank, July 2007), http://www-wds.worldbank.org/external/default/WDSContentServer/IW3P/IB/2007/07/31/000158349_20070731143402/Rendered/PDF/wps4293.pdf (accessed April 6, 2009).

[xcv] Helmy M. Eid, Samia M. El-Marsafawy, and Samiha A. Ouda, "Assessing the Economic Impacts of Climate Change on Agriculture in Egypt: A Ricardian Approach" Policy Research Working Paper (Development Research Group of The World Bank, July 2007), http://www-wds.worldbank.org/external/default/WDSContentServer/IW3P/IB/2007/07/31/000158349_20070731143402/Rendered/PDF/wps4293.pdf (accessed April 6, 2009).

[xcvi] Ragab Ragab, e-mail message to author, May 27, 2009.

[xcvii] R. Schubert, H.J. Schellnhuber, N. Buchmann, A. Epiney, R. GrieBhammer, M. Kulessa, D. Messner, S. Rahmstorf, J. Schmid. *Climate Change as a Security Risk*, German Advisory Council on Global Change (WBGU), trans. Christopher Hay and Seeheim-Jugenheim (London and Sterling, VA: Earthscan, 2008).

[xcviii] Ayman F. Abou-Hadid, "Assessment of Impacts, Adaptation, and Vulnerability to Climate Change in North Africa: Food Production and Water Resources" A Final Report Submitted to Assessments of Impacts and Adaptations to Climate Change (AIACC), Project No. AF 90 (Washington, D.C.: International START Secretariat, 2006).

[xcix] R. Schubert, H.J. Schellnhuber, N. Buchmann, A. Epiney, R. GrieBhammer, M. Kulessa, D. Messner, S. Rahmstorf, J. Schmid. *Climate Change as a Security Risk*, German Advisory Council on Global Change (WBGU), trans. Christopher Hay and Seeheim-Jugenheim (London and Sterling, VA: Earthscan, 2008).

[c] World Bank, World Development Indicators Online (accessed April 15, 2009).

[ci] Helmy M. Eid, Samia M. El-Marsafawy, and Samiha A. Ouda, "Assessing the Economic Impacts of Climate Change on Agriculture in Egypt: A Ricardian Approach" Policy Research Working Paper (Development Research Group of The World Bank, July 2007), http://www-wds.worldbank.org/external/default/WDSContentServer/IW3P/IB/2007/07/31/000158349_20070731143402/Rendered/PDF/wps4293.pdf (accessed April 6, 2009).

[cii] World Bank, "The World Bank Middle East and North Africa Region (MENA) Sustainable Development Sector Department (MNSSD) Regional Business Strategy to Address Climate Change" (Preliminary draft for consultation and feedback, World Bank, Washington, DC, November 2007), http://siteresources.worldbank.org/INTCLIMATECHANGE/Resources/MENA_CC_Business_Strategy_Nov_2007_Revised.pdf (accessed April 15, 2009); and R. Schubert, H.J. Schellnhuber, N. Buchmann, A. Epiney, R. GrieBhammer, M. Kulessa, D. Messner, S. Rahmstorf, J. Schmid. *Climate Change as a Security Risk*, German Advisory Council on Global Change (WBGU), trans. Christopher Hay and Seeheim-Jugenheim (London and Sterling, VA: Earthscan, 2008).

[ciii] UNU-INWEH (United Nations University's Canadian-based International Network on Water, Environment and Health), "Experts Advice World Policies to Cope with Causes, Rising Consequences of Creeping Desertification: If Unaddressed, Experts Warn Waves of Environmental Refugees may Follow; Effective Responses to Desertification will also Mitigate Climate Change" (Joint International Conference, Algiers, December 14, 2006), http://www.inweh.unu.edu/inweh/drylands/Algiers_news_release-Final.pdf (accessed April 15, 2009).

[civ] Nigel W. Arnell, "Climate change and global water resources: SRES emissions and socio-economic scenarios." *Global Environmental Change* 14 (2004): 31–52.

[cv] World Bank, "The World Bank Middle East and North Africa Region (MENA) Sustainable Development Sector Department (MNSSD) Regional Business Strategy to Address Climate Change" (Preliminary draft for consultation and feedback, World Bank, Washington, DC, November 2007), http://siteresources.worldbank.org/INTCLIMATECHANGE/Resources/MENA_CC_Business_Strategy_Nov_2007_Revised.pdf (accessed April 15, 2009).

[cvi] M. Benassi, "Drought and climate change in Morocco. Analysis of precipitation field and water supply," (*Options Méditerranéennes* Series A, no. 80, 2008) in *Drought management: scientific and technological innovations*, ed. A. Lopez-Francos, 83-86 (Zaragoza, Spain: CIHEAM-IAMZ, 2008).

[cvii] Tom Pfeiffer, "Climate Change Threatens North Africa Food Supply," *Reuters*, June 27, 2007, http://www.planetark.com/dailynewsstory.cfm/newsid/42814/story.htm (accessed May 7, 2009).

[cviii] Ayman F. Abou-Hadid, "Assessment of Impacts, Adaptation, and Vulnerability to Climate Change in North Africa: Food Production and Water Resources" A Final Report Submitted to Assessments of Impacts and Adaptations to Climate Change (AIACC), Project No. AF 90 (Washington, D.C.: International START Secretariat, 2006).

[cix] FAONewsroom, Agriculture in the Near East likely to suffer from climate change: The hungry and poor will be most affected – FAO meeting debates impact on the region, March 3, 2008, Rome/Cairo, http://www.fao.org/newsroom/en/news/2008/1000800/index.html.

[cx] Abigail Somma, "Squeezing the Most out of Scarce Water Resources,"*IFPRI Forum*, I (2009).

[cxi] Maarten de Wit and Jacek Stankiewicz, "Changes in Surface Water Supply Across Africa with Predicted Climate Change." *Science* 311, no. 5769 (2006): 1917-1921.

[cxii] Elisabeth Meze-Hausken, "Migration Caused By Climate Change: How Vulnerable Are People in Dryland Areas?" *Mitigation and Adaptation Strategies for Global Change* 5 (2000), 379-406.

[cxiii] FMO (Forced Migration Online), "Africa," http://www forcedmigration.org/browse/regional/africa.htm (accessed April 27, 2009).

[cxiv] Hein de Haas, "Trans-Saharan Migration to North Africa and the EU: Historical Roots and Current Trends," (Migration Information Source, November 2006), http://www.migrationinformation.org/feature/display.cfm?ID=484 (accessed April 27, 2009).

[cxv] FMO (Forced Migration Online), "Africa," http://www forcedmigration.org/browse/regional/africa.htm (accessed April 27, 2009); and Hein de Haas, "Trans-Saharan Migration to North Africa and the EU: Historical Roots and Current Trends," (Migration Information Source, November 2006), http://www migrationinformation.org/feature/display.cfm?ID=484 (accessed April 27, 2009).

[cxvi] Hein de Haas, "Trans-Saharan Migration to North Africa and the EU: Historical Roots and Current Trends," (Migration Information Source, November 2006), http://www.migrationinformation.org/feature/display.cfm?ID=484 (accessed April 27, 2009).

[cxvii] R. Schubert, H.J. Schellnhuber, N. Buchmann, A. Epiney, R. GrieBhammer, M. Kulessa, D. Messner, S. Rahmstorf, J. Schmid. *Climate Change as a Security Risk*, German Advisory Council on Global Change (WBGU), trans. Christopher Hay and Seeheim-Jugenheim (London and Sterling, VA: Earthscan, 2008).

[cxviii] Hein de Haas, "Trans-Saharan Migration to North Africa and the EU: Historical Roots and Current Trends," (Migration Information Source, November 2006), http://www.migrationinformation.org/feature/display.cfm?ID=484 (accessed April 27, 2009).

[cxix] United Nations Department of Economic and Social Affairs, cited in R. Schubert, H.J. Schellnhuber, N. Buchmann, A. Epiney, R. GrieBhammer, M. Kulessa, D. Messner, S. Rahmstorf, J. Schmid. *Climate Change as a Security Risk*, German Advisory Council on Global Change (WBGU), trans. Christopher Hay and Seeheim-Jugenheim (London and Sterling, VA: Earthscan, 2008).

[cxx] Hein de Haas, "Trans-Saharan Migration to North Africa and the EU: Historical Roots and Current Trends," (Migration Information Source, November 2006), http://www.migrationinformation.org/feature/display.cfm?ID=484 (accessed April 27, 2009).

[cxxi] IPCC [Intergovernmental Panel on Climate Change]. *Climate Change 2007: Impacts, Adaptation and Vulnerability*, eds. Martin Parry, Osvaldo Canziani, Jean Palutikof, Paul van der Linden, and Clair Hanson (Cambridge: Cambridge University Press, 2007): 435.

[cxxii] IPCC [Intergovernmental Panel on Climate Change]. *Climate Change 2007: Impacts, Adaptation and Vulnerability*, eds. Martin Parry, Osvaldo Canziani, Jean Palutikof, Paul van der Linden, and Clair Hanson (Cambridge: Cambridge University Press, 2007): 435.

[cxxiii] German Advisory Council on Global Change (WBGU), *Climate Change as a Security Risk*, eds. R. Schubert, H.J. Schellnhuber, N. Buchmann, A. Epiney, R. Grießhammer, M. Kulessa, D. Messner, S. Rahmstorf, and J. Schmid (London: Earthscan , 2008).

[cxxiv] BBC, "Africa militants make UK threat," *BBC*, April 27, 2009, http://news.bbc.co.uk/2/hi/africa/8020504.stm (accessed April 30, 2009).

[cxxv] German Advisory Council on Global Change (WBGU), *Climate Change as a Security Risk*, eds. R. Schubert, H.J. Schellnhuber, N. Buchmann, A. Epiney, R. Grießhammer, M. Kulessa, D. Messner, S. Rahmstorf, and J. Schmid (London: Earthscan , 2008).

[cxxvi] Elisabeth Meze-Hausken, "Migration Caused By Climate Change: How Vulnerable Are People in Dryland Areas?" *Mitigation and Adaptation Strategies for Global Change* 5 (2000), 379-406.

[cxxvii] Robert J. Nicholls, Frank M.J. Hoozemans and Marcel Marchand, "Increasing flood risk and wetland losses due to global sea-level rise: regional and global analyses" *Global Environmental Change* 9, supplement 1 (1999): S69-S87; and Robert J. Nicholls and Richard S.J. Tol, "Impacts and Responses to sea-level rise: a global analysis of the SRES scenarios over the twenty-first century" *Phil. Trans. R. Soc. A* 364, no. 1841 (2006): 1073-1095.

[cxxviii] IPCC [Intergovernmental Panel on Climate Change]. *Climate Change 2007: Impacts, Adaptation and Vulnerability*, eds. Martin Parry, Osvaldo Canziani, Jean Palutikof, Paul van der Linden, and Clair Hanson (Cambridge: Cambridge University Press, 2007).

[cxxix] For a review of these studies, see Mohamed El Raey, "Adaptation to climate change for sustainable development in the coastal zone of Egypt," Global Forum on Sustainable Development (Organization for Economic

Co-operation and Development, February 7, 2005), http://www.oecd.org/dataoecd/37/21/34692998.pdf (accessed May 6, 2009).

[cxxx] Susmita Dasgupta, Benoit Laplante, Craig Meisner, David Wheeler and Jianping Yan, "The Impact of Sea Level Rise on Developing Countries: A Comparative Analysis," World Bank Policy Research Paper 4136 (World Bank, February 2007), http://www-wds.worldbank.org/servlet/WDSContentServer/WDSP/IB/2007/02/09/000016406_20070209161430/Rendered/PDF/wps4136.pdf (assessed May 6, 2009).

[cxxxi] Susmita Dasgupta, Benoit Laplante, Craig Meisner, David Wheeler and Jianping Yan, "The Impact of Sea Level Rise on Developing Countries: A Comparative Analysis," World Bank Policy Research Paper 4136 (World Bank, February 2007), http://www-wds.worldbank.org/servlet/WDSContentServer/WDSP/IB/2007/02/09/000016406_20070209161430/Rendered/PDF/wps4136.pdf (assessed May 6, 2009); and FAO (Food and Agriculture Organization of the United Nations), "World agriculture: towards 2015/2030 Summary report" (Rome: 2002), ftp://ftp fao.org/docrep/fao/004/y3557e/y3557e.pdf (accessed May 6, 2009).

[cxxxii] Susmita Dasgupta, Benoit Laplante, Craig Meisner, David Wheeler and Jianping Yan, "The Impact of Sea Level Rise on Developing Countries: A Comparative Analysis," World Bank Policy Research Paper 4136 (World Bank, February 2007), http://www-wds.worldbank.org/servlet/WDSContentServer/WDSP/IB/2007/02/09/000016406_20070209161430/Rendered/PDF/wps4136.pdf (assessed May 6, 2009).

[cxxxiii] World Bank, "The World Bank Middle East and North Africa Region (MENA) Sustainable Development Sector Department (MNSSD) Regional Business Strategy to Address Climate Change" (Preliminary draft for consultation and feedback, World Bank, Washington, DC, November 2007), http://siteresources.worldbank.org/INTCLIMATECHANGE/Resources/MENA_CC_Business_Strategy_Nov_2007_Revised.pdf (accessed April 15, 2009).

[cxxxiv] World Travel and Tourism Council, s.v.v. "Egypt," "Libya," "Tunisia," "Algeria," and "Morocco" http://www.wttc.org/eng/Tourism_Research/Tourism_Economic_Research/Country_Reports/ (accessed April 26, 2009).

[cxxxv] World Bank, "Climate Change and Environment & Natural Resources: Overview Analytical and Advisory Activities," http://go.worldbank.org/TZ3SBTWB30 (accessed April 26, 2009).

[cxxxvi] D. Moran "Climate Change and Regional Security" (workshop, Department of National Security Affairs at the Naval Postgraduate School, sponsored by Long Range Assessment Unit of the National Intelligence Council, Monterey, California, December 11-13, 2007).

[cxxxvii] Waleed Hazbun, "The Development of Tourism Industries in the Arab World: Trapped Between the Forces of Economic Globalization and Cultural Commodification" (30th Annual Convention of the Association of Arab-American University Graduates (AAUG), Washington, DC. November 1, 1997), http://hazbun mwoodward.com/Globalization.html (accessed April 26, 2009).

[cxxxviii] D. Moran "Climate Change and Regional Security" (workshop, Department of National Security Affairs at the Naval Postgraduate School, sponsored by Long Range Assessment Unit of the National Intelligence Council, Monterey, California, December 11-13, 2007).

[cxxxix] Center for Strategic and International Studies, "The Politics of North African Energy," Maghreb Roundtable (Washington, D.C.: Middle East Program at the Center for Strategic and International Studies, March 2006).

[cxl] Energy Information Administration (EIA), "International Energy Outlook 2008: Chapter 3 – Natural Gas" (Report # DOE/EIA-0484, June 2008), http://www.eia.doe.gov/oiaf/ieo/nat_gas html (accessed May 7, 2009).

[cxli] Energy Information Administration (EIA), "Libya: Background" July 2007, http://www.eia.doe.gov/cabs/Libya/Full html (accessed May 7, 2009).

[cxlii] Energy Information Administration (EIA), "Libya: Background" July 2007, http://www.eia.doe.gov/cabs/Libya/Full html (accessed May 7, 2009).

[cxliii] Energy Information Administration (EIA), "Algeria: Background" March 2007, http://www.eia.doe.gov/cabs/Libya/Full html (accessed May 8, 2009).

[cxliv] Hans Günter Brauch, "Energy interdependence in the western Mediterranean" *Mediterranean Politics* 1, no. 3 (1996): 295-319.

[cxlv] Energy Information Administration (EIA), "Libya: Background" July 2007, http://www.eia.doe.gov/cabs/Libya/Full html (accessed May 7, 2009); and Central Intelligence Agency (CIA), *The World Factbook*, https://www.cia.gov/library/publiscations/the-world-factbook/index html (accessed May, 6 2009).

53

cxlvi Energy Information Administration (EIA), "Libya: Background" July 2007, http://www.eia.doe.gov/cabs/Libya/Full html (accessed May 7, 2009); and Central Intelligence Agency (CIA), *The World Factbook*, https://www.cia.gov/library/publiscations/the-world-factbook/index html (accessed May, 6 2009).
cxlvii Energy Information Administration (EIA), "Algeria: Background" March 2007, http://www.eia.doe.gov/cabs/Libya/Full html (accessed May 8, 2009).
cxlviii Energy Information Administration (EIA), "Libya: Background" July 2007, http://www.eia.doe.gov/cabs/Libya/Full html (accessed May 7, 2009).
cxlix Hans Günter Brauch, "Energy interdependence in the western Mediterranean" *Mediterranean Politics* 1, no. 3 (1996): 295-319.
cl Plan Bleu Regional Activity Center "Climate Change and Energy in the Mediterranean," (July 2008), http://www.eib.org/attachments/country/femip-study-climate-change-and-energy-in-the-mediterranean.pdf (accessed May 8, 2009).
cli Intergovernmental Panel on Climate Change, "Glossary," *Climate Change 2007: Impacts, Adaptation and Vulnerability: Contribution of Working Group II to the Fourth Assessment Report of the Intergovernmental Panel on Climate Change*, eds. M.L. Parry, O.F. Canziani, J.P. Palutikof, P.J. van der Linden and C.E. Hanson (Cambridge: Cambridge University Press 2007).
clii G. Yohe and R. Tol, "Indicators for social and economic coping capacity—moving toward a working definition of adaptive capacity," *Global Environmental Change* 12 (2002): 25-40
cliii E.L. Malone and A.L. Brenkert, "Vulnerability, sensitivity, and coping/adaptive capacity worldwide," *The Distributional Effects of Climate Change: Social and Economic Implications*, M. Ruth and M. Ibarraran, eds., Elsevier Science, Dordrecht (in press).
cliv N. Nakicenovic and R. Swart, *Special Report on Emissions Scenarios* (Cambridge: Cambridge University Press 2000).

This page intentionally left blank.

SPECIAL REPORT

NORTH AFRICA:
THE IMPACT OF CLIMATE CHANGE TO 2030
(SELECTED COUNTRIES)

A COMMISSIONED RESEARCH REPORT

www.ingramcontent.com/pod-product-compliance
Lightning Source LLC
Chambersburg PA
CBHW081258180526
45170CB00007B/2478